AMERICAN

DESIGN CLASSICS

AMERICAN

DESIGN CLASSICS

By Nada Westerman and Joan Wessel

Introduction by Arthur J. Pulos

DESIGN PUBLICATIONS

Editor Randolph McAusland
Design Fulton + Partners
Design Director AnnLee Polus
Designer Clive Jacobson
Photography Joseph Kugielsky
Chairman Design Publications James F. Fulton
Typesetting David Seham, Inc.

Library of Congress Cataloging in Publication Data

Library of Congress Catalog
Card Number 85-062059

Wessel, Joan

Westerman, Nada

American Design Classics

Bibliography: p. 190

I. Product design

II. Industrial design

III. History of American design

ISBN 0-934341-00-1

A Design Publications Book

Prepared, produced and first published in 1985 by Design Publications, Inc., 330 West 42nd Street, New York, NY 10036.

CONTENTS

1. Technological/innovation.

New product directions inspired by technological advances.

2. Function/form.

The functional requirements of the product are resolved in an expressive form.

3. Form

Practical objects are translated into sculptural forms.

4. Materials.

Appropriate forms designed from new materials: The innovative use of materials.

5. Human Factors.

Compatibility of the product to the user is given primary consideration.

Preface

Because we have come to depend upon our "designed" environments—at home, at work and at play—the designer of business and consumer products has a powerful impact on the way we live our lives. Why products look and work the way they do, however, remains a mystery to the public at large. Industrial designers are the "they" that we refer to in our colloquial manner when we say, "They sure made this stereo confusing" or "They really improved this stove." But who are "they" and how do they work? "Considering that designs are everywhere," says Ralph Caplan in *By Design,* "what designers do is neither highly visible nor widely understood. We are all aware of the results but not of the activity."

We believed that the public could understand the elements of good design through an in-depth examination of how great designs evolve. For students, a study of the design process would illuminate the problem-solving skills employed in successful design solutions.

As our research progressed, we found that contemporary American design was severely neglected; worse, it was often dismissed as unworthy. While European and Japanese designs had been the subject of exhibitions and media exposure, and American designs from the 1930s to the 1950s enjoyed a resurgence of curatorial interest, American design of the past twenty-five years remained unexamined. This period, therefore, seemed to be an apt subject for study and a fresh approach.

More directly than other art forms, industrial design reflects changing social and cultural values. The 1960s were watershed years in the U.S., for new directions were charted then that endure today. Through an exhibition of consumer products from this period, we could explore the impact of these new ideas on American design.

In selecting the designs for the exhibition, we sought a wide geographic representation and a sense of the diversity of the design practice, including corporate design departments, consultant design groups, the designer-craftsman and the environmental designer.

An interesting design history, and the availability of documentation were also considerations. It was our intention, in following the step-by-step development of each design, to explore the designer's approach to the design problem. We hoped to learn what questions he asked, what decisions he made, and the nature of his collaboration with the marketing department and engineers. We wanted to exhibit this process visually through sketches, drawings, molds, models, blueprints, photographs, and video tapes.

We also sought to balance the well-known "classics" of contemporary American design—the Trimline Telephone, the IBM Selectric typewriter and the Eames Aluminum Group Chair, for example—with more recent designs, such as Eureka's Mighty Mite vacuum cleaner, the Magnavox Compact Disc Player and the Montare Mountain Bike.

After compiling a list of two hundred products, we met with a panel of distinguished designers and critics. And it was Ralph Caplan, Neils Diffrient, Richard Penney, John Pile and Arnold Wolf who helped guide our final selections. Based on personal interviews and responses to our detailed questionnaire, we were then able to develop case histories for each project.

By concentrating on consumer products, we realized that only a small sampling of industrial design activity could be shown. To provide a context and to illustrate the work of designers in other fields (graphics, packaging, environments, space planning, systems, industrial and medical products), Rosalie Dolmatch and Katherine Moore produced an audio-visual program on the history and scope of American industrial design to serve as an orientation for the exhibition.

The consumer products of the last twenty-five years reflect the designer's response to changing values, technology and lifestyles. Concern with the individual's needs has stimulated designers to consider human factors and design for the aged, the handicapped and the disabled. Rapidly changing technology has required designers to become involved in bridging the gap between man and the machine, as "user friendly" has entered our vocabulary. As the work force shifts from factory to the automated office, designers are developing systems that will accomodate the most advanced technological equipment with the least stress for the user. These products reflect changes in family and household structure and changing lifestyles. The busy and often fragmented family, the serious hobbyist, the physical fitness enthusiast, and the gourmet cook are well served by contemporary designers.

For a better understanding of the design process, products were divided into five categories reflecting the main issues in each design solution: 1. Technological Innovation (new product directions inspired by technological advances), 2. Form/Function (funcational requirements of the product resolved in an expressive form), 3. Form (practical objects translated into sculptural forms), 4. Materials (appropriate forms designed from new materials and the innovative use of new materials), 5. Human Factors (compatibility of the product to the user given primary consideration).

The Product of Design offers an unusual opportunity to explore the creative process: mental and tangible steps behind the well-designed consumer product. It also measures the designer's ability to synthesize the often conflicting product requirements: function, aesthetics, human factors, materials, manufacturing, marketing and cost. The exhibition and book provide a fresh look at U.S. design, exploding the myth that good design cannot survive in the American mass market. The best designers work creatively within the system and respond to the challenge by overcoming constraints to create effective, innovative design solutions.

Joan Wessel, Nada Westerman

Acknowledgements

American Design Classics is the publishing expression of The Product of Design exhibition originated by The Katonah Gallery, Katonah, New York, in 1984. Mary Barnes, a designer and member of the gallery advisory board, suggested the exhibition and recommended that the focus be on the design *process*. Once we decided to concentrate on process, and not merely objects, we set out to explore how designs evolve in practice. This led us to design schools, design studios, and the design departments of major corporations. At every level, we received invaluable assistance, and we want to acknowledge and thank all those who have contributed to the exhibition and to the book.

Our initial research took us back to school. To The University of Bridgeport and Robert Redmann, to Jack Aurelli at Pratt Institute, and to Marc Harrison at the Rhode Island School of Design. We met with faculty and students, reviewed the fundamentals of design, and were introduced to the scope of formal design education. Once the focus was established, we began to select a group of products that would reflect the best in contemporary American product design. We studied back issues of *ID* magazine for outstanding designs, especially the *Annual ID Design Review.* We consulted with David McFadden at the Cooper-Hewitt Museum, Stewart Johnson at The Museum of Modern Art, Christopher Wilk of The Brooklyn Museum, Derek Ostergaard, a museum consultant, and with the education department of The Philadelphia Museum of Art. The Industrial Designers' Society of America shared its resources, specifically, information on its Industrial Design Excellence Award recipients. Members of the New York chapter—especially Andrew Serbinski, Larry Porcelli, Richard Penney and Bonnie Srolowitz—were helpful consultants. We met with hundreds of designers from all parts of America. We are indebted to John Pile, the project consultant, and to Dagmar Arnold, Morison Cousins, John Graham, Walter Kraus, Charles Mauro, Eudora Moore, William Lansing Plumb, Arthur Pulos, Deane Richardson, Allen Samuels, RitaSue Siegel, Raymond Spilman, and Cooper Woodring, all of whom gave unselfishly of their time and expertise.

We are deeply grateful to the manufacturers whose products were loaned for exhibition and for the cooperation of the designers who responded to our questionnaire, searched their files for documentation, and interrupted busy schedules to meet our deadlines. Herman Miller, Inc., generously provided photographs from their archives for a photo essay on George Nelson and Charles Eames, and a selection of Eames films that were shown at The Gallery in the course of the exhibition. Our thanks also to Don Rorke, of Knoll International, whose foresight enabled the exhibition to become part of Worldesign 85 in Washington, D.C., and to the Gallery Association of New York State for managing the exhibition's tour.

A process exhibition containing so many diverse elements required a first rate exhibition design. Murry Gelberg's solution, using an industrial landscape, achieved the balance needed between the visual and interpretive. Andrea Mauro provided the promotional graphics, and Carolyn Bucha executed the illustrative labelling materials.

Support for the exhibition was provided by public funds from the New York Council on the Arts and the Westchester Arts Fund of the Council for the Arts in Westchester, supported by corporate contributions and the County of Westchester. Additional funds were received from The Heckscher Foundation for Children, the IBM Fund for Community Service and Exxon Corporation. We also wish to thank the staff of The Katonah Gallery for their unstinting support, and Randolph McAusland and Design Publications, Inc., for this opportunity to translate the exhibition into a book so that the ideas and images could reach a wider audience.

J.W., N.W.

Introduction

Industrial design is the object of increasing public curiosity in the United States. There is something about the integration of design and technology, symbolized by the space program, that seems to be uniquely American. There are even some who believe that American children are born with an instinct for design. Infants in cribs play with "busy boxes"; toddlers build with blocks, then graduate to Legos. Arts and crafts are a staple of American school programs. Later, after a brief romance with video games, an impressive number of teenagers develop into precocious computer hackers. Tens of thousands of high school students then move on to college to become engineers and scientists. Only a few hundred each year opt for industrial design. Yet this unique and highly demanding discipline is rich with history and exercises an enormous influence on the way we live.

From the beginning, industrial designers have dedicated their imagination and sensitivity to the development of products that serve the physical needs and psychological desires of the public. They have been motivated by a conviction that a man-made object acquires beauty to the degree that it approaches the ideal *typeform* of its species. Benjamin Franklin stated the American designer's creed "Nothing is good or beautiful but in the measure that it is useful."

Industrial design, as a creative synthesis of the cultural and technological aspects of manufactured products, emerged as a distinct profession in the U.S. in the 1920s. Some believe that it was stimulated by the Modernist furniture and accessories that captured so much attention at the 1925 International Exposition of Modern Decorative and Industrial Arts in Paris. An even stronger motivation may have been the fact that the U.S. refused an invitation to participate because Secretary of Commerce Herbert Hoover concluded, after consultation with manufacturers, that Americans had nothing original to show. The seminal year for industrial design may be 1927. It was the year that exhibitions from the Paris Exposition were shown in America, and it was the year of automobile model changes. By 1930, a number of illustrators, as well as architects, decorators, engineers, and stage set designers—both American and foreign—had been drawn into the new profession, attracted by the dramatic opportunities to design products for industry.

Even during the Depression American optimism was expressed by the soaring new skyscrapers of New York City, the Century of Progress Exposition in Chicago, all-metal airplanes, Buckminster Fuller's Dymaxion and Chrysler's Airflow automobiles. Streamlined trains gave form and a name to an era that promised better times ahead for those who had faith in American industry. It became evident that a cautious public was willing to spend its limited dollars on manufactured products if they offered improved function and an entirely new appearance. "In the Thirties," Edgar Kaufmann later wrote, "it seemed to many of us that the period's design vitality was centered in America, rather than Europe, perhaps for the first time."

In the immediate post-war era, manufacturers and their designers rode the tide of a pent-up public demand for products and a higher standard of living; at the same time, museums and department stores did their best to steer the public toward "good design." As markets became saturated, however, manufacturers and designers resorted to artificial obsolescence and baroque forms to stimulate further consumption. The consequence was a wave of public resistance, even protest, that introduced a period of design maturity. American products of the past twenty-five years, as a result, have become better looking and more considerate of the consumer.

Industrial design is a profession for those whose minds are tuned to the future even though their hearts may be in the past and their feet planted in the present. Industrial designers seem to have several qualities in common: a sensitivity to the man-made world with a demonstrable empathy for their fellow humans and a natural affinity for cultural values; an innate curiosity that drives them to wonder how things are made, what makes them work, and why they must be as they are, and, finally, a compulsive ingenuity and a passion for imaginative solutions that meet their own criteria for excellence.

It may be suggested, at the risk of oversimplification, that there are two approaches to the practice of design: private and public. Private designers, such as the late Charles Eames, are essentially preoccupied with solving problems of their own choosing, and they devote their time and talent to exploring concepts outside the mainstream. Most industrial designers, however, fall into the public category. Though they may savor private projects tucked away in their minds, for the most part, they undertake assignments brought to them by clients. Eliot Noyes was such a designer.

In everyday practice, the design process has much in common with other creative service activities. It follows well-worn paths of acquisition and analysis of pertinent information from which a set of parameters for problem-solving are established. Alternative solutions are considered, the best being prepared for review by the relevant parties. The most promising direction is developed into mechanical drawings, models and other aids to manufacture. In most cases full-scale prototypes are built and tested before final production. The final step is an evaluation of the product's performance in the field and its appeal to consumers.

At the heart of the American design process is the realization that mass-produced products are essentially democratic in character since they must be tailored to fit the broadest possible logical market. Such products are fundamentally transient, destined to disappear humbly after their period of service. The designer's special challenge is to extract the most from the energy and materials that are to be used in production and, in the process, to reduce the cost of the object without jeopardizing the quality of its service. In other words, to do more with less.

Designers are aware that every product is in a constant state of evolution—moving from its original invented form through a series of rises and plateaus in its search for perfection. A rise occurs when new technology or shifts in living and working patterns makes changes necessary. A plateau occurs when technology and

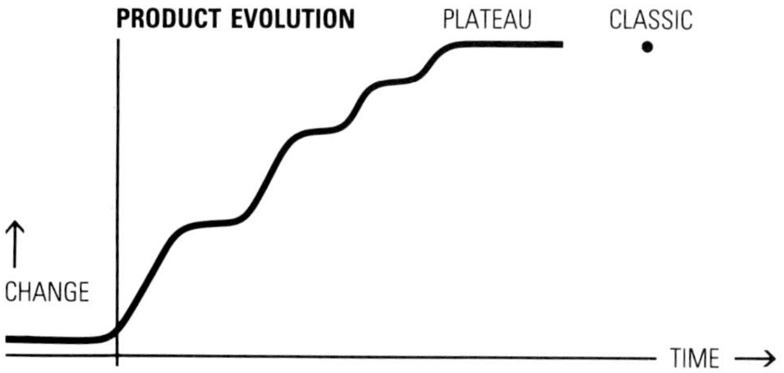

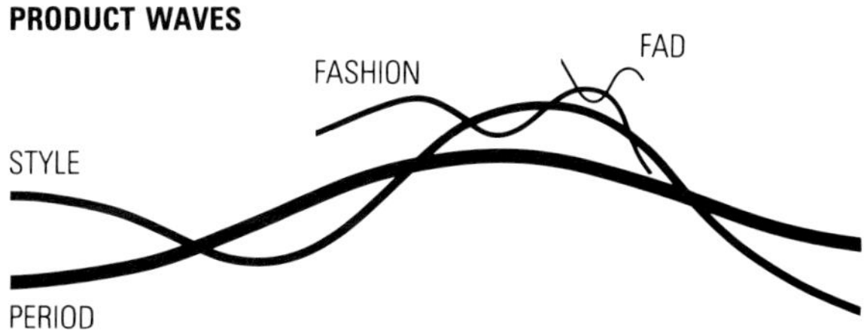

public expectation are in equilibrium; however, public pressure and competition in a static market may lead to adoption of styling mannerisms that *suggest* progress is being made. Such pretenses are not always a "positive evil" as Horatio Greenough observed, because they inevitably force another upward advance. On occasion a product may be displaced by another that provides the same service in a better way. In that event the original product will be abandoned. Should it be remembered with affection, however, it may be brought back to be admired as a work of art—a "classic" absolved of functional responsibility. Witness the current interest in Model T's, Victrolas and the reproduction of oak ice boxes, "Candlestick" telephones and "Gothic" radios.

Designers are sensitive to the fact that every product that man makes is afloat on a history of expressive periods that rise and fall in great swells on the surface of a cultural ocean. While designers cannot alter significantly the aesthetic character of their period, they are able to influence the waves of style that more closely match their own life spans. Fashions are aberrations that swirl around the waves of style on the annual or seasonal pattern, while fads are those fleeting fancies of expression that appear like foam on the crest of the wave and are then gone with the next puff of desire.

American Design Classics is simply a manufactured response to some human need or desire. It is bounded by the three dimensions of industrial design: human (physical and psychological considerations), aesthetic (desires and aspirations), and technological (potential and limitations). In turn, the trigon is contained within marketing—or social relevance. Outside the circle of marketing are those factors that control the whole: industry that invests its resources in hope that its product will sell; a public that reserves the right to accept or reject it, and, finally, a government that mediates between industry and the public and is responsible to the whole. A manufactured product is only a small part of a larger system. Its value, however, lies in the fact that it can provide thousands with a service that they might otherwise be denied.

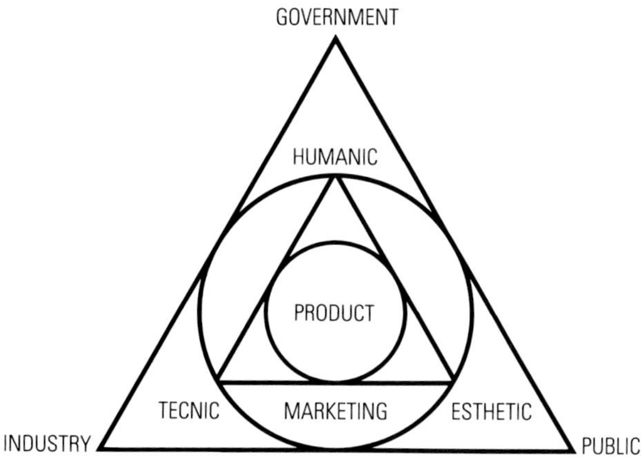

As the humane and aesthetic conscience of industry, industrial design today is concerned with the transformation of scientific advances into innovative products that correlate with rapidly changing personal and social patterns. American designers and their clients are constantly probing the frontiers of technology for products and services that may have no precedent but will shape tomorrow's man-made environment. They realize that as an increasing percentage of products sold in this country are manufactured abroad, useful applications must be found for America's ever more exotic research achievements.

Innovative designers are participating in the technological and electronic revolutions of today. By the end of this century one may expect that the man-made environment will have been transformed as dramatically as it was at the beginning of the century, when electricity and fossil fuel engines first came into general use. Our present preoccupation with recycling the past, with post-modernist architectural forms and aberrant fashions and fads in furnishings may be taken as clearing the way for a new design spirit that seeks a fresh fusion between science, art and society. There is increasing evidence that the man-made environment is moving toward a technosphere in which mankind may live comfortably and safely, independent of the threatened natural resources of this planet.

Arthur J. Pulos, FIDSA

AMERICAN

1
TECHNOLOGICAL INNOVATION

DESIGN CLASSICS

Eames® Aluminum Group Lounge Chair 1958

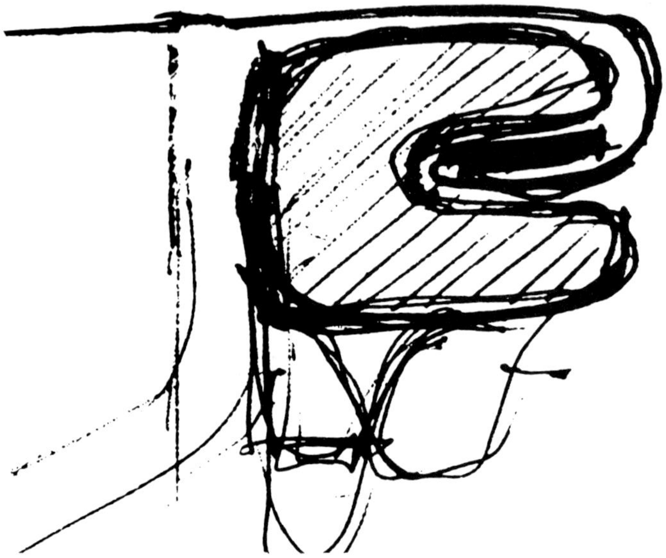

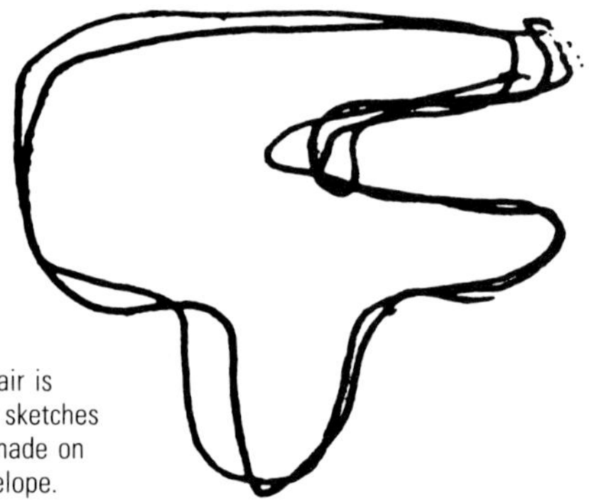

The idea for the chair is found in these first sketches by Charles Eames made on the back of an envelope.

The Design Problem: The catalyst for the design program that led to the development of the Aluminum Group was a need for quality outdoor furniture.

The cross section, which Eames recalled drawing on the back of an envelope, was the actual idea for the chair—"the device that made it possible." The cross section was "a system of connections based on tensions which serve to support the body." Fabric, inserted into a narrow slot and then brought around, is held with great force when in tension. The force increases with resistance.

The seat and back are one continuous plane slung between structural ribs of die-cast aluminum. Base and back stretchers keep sides apart, also holding the fabric in tension. Stiffening strips, hidden in the fabric folds at strategic intervals, support the body and prevent the sag or ordinary sling chairs.

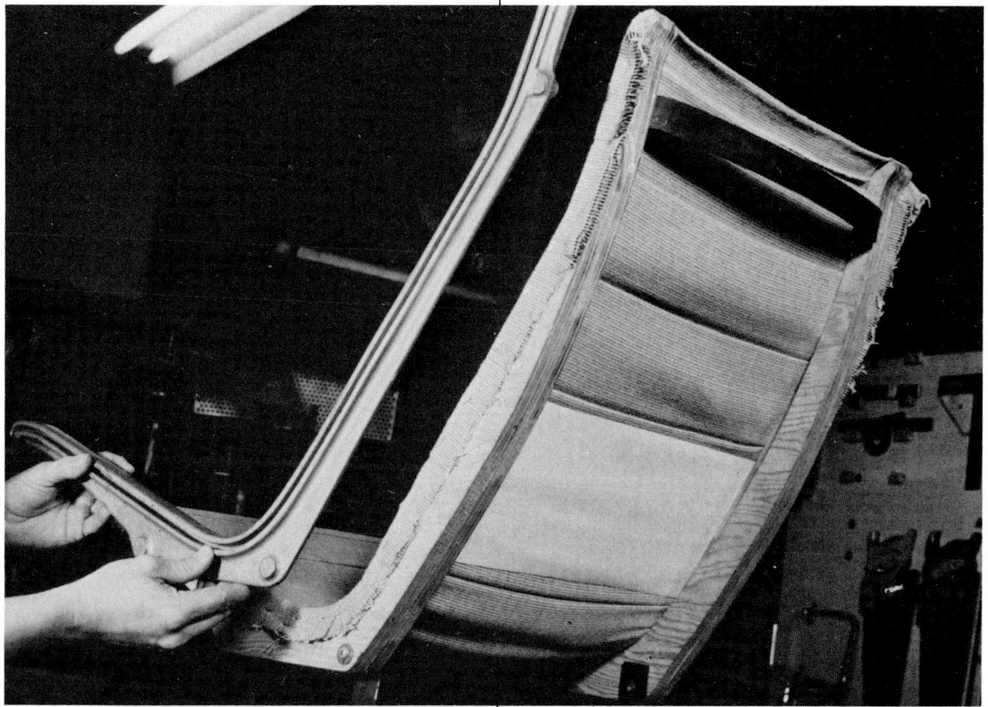

Stiffening strips.

Back stretcher.

Designer Charles Eames
Manufacturer Herman Miller, Inc.

Eames® Aluminium Group Lounge Chair 1958

Now in use at major airports and in other public buildings, tandem seating evolved from the Aluminum Group. Comfort, durability, ease of maintenance and quick, low-cost parts replacement make this an ideal design for public seating. Seats are secured to a continuous floating steel T-beam so that large spans are possible.

The Eames Aluminum Group Side Chair is in the Design Collection of the Museum of Modern Art.

Charles Eames in his studio with experimental forms for the chair base.

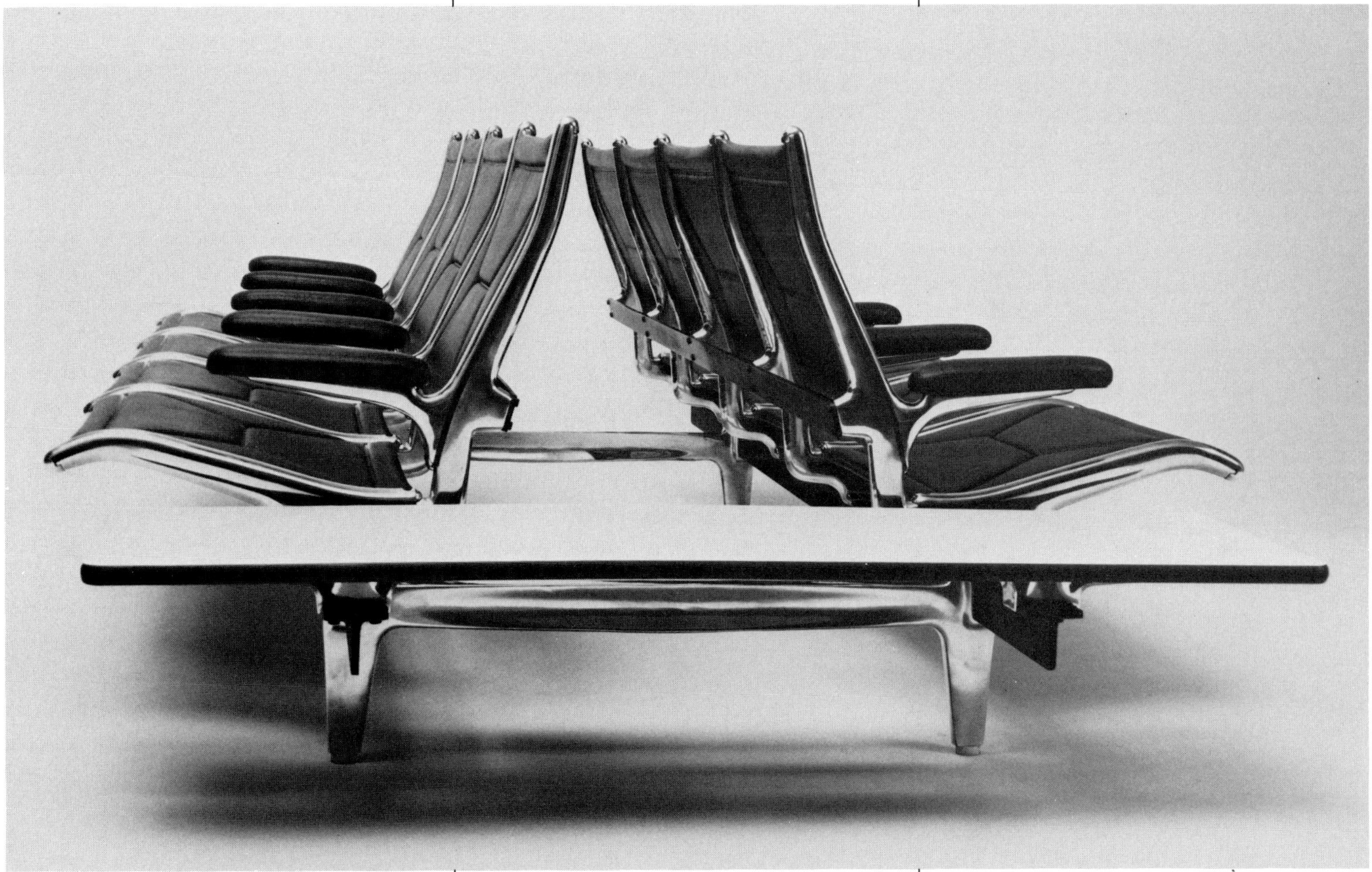

Eames Aluminum Group Lounger.

IBM Selectric I 1961

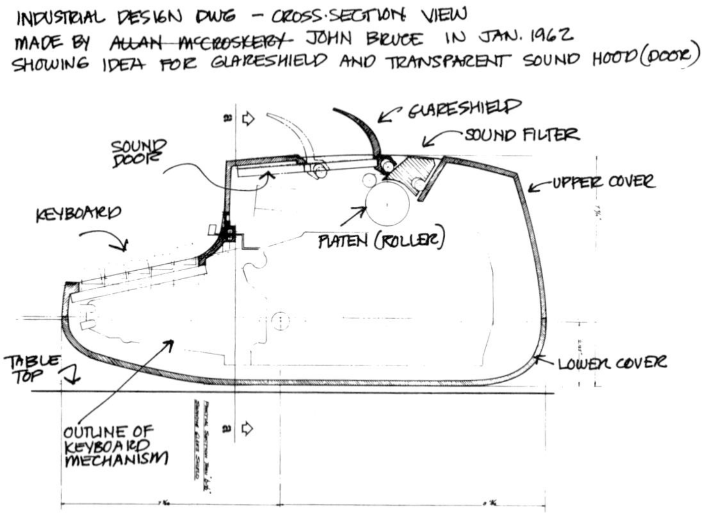

The Design Problem: To apply new principles to a new form of typewriter and to eliminate all distracting details and the multiple parts common to other machines.

The Design History: T.J. Watson, Jr., IBM chairman, asked Eliot Noyes to undertake industrial design consulting for IBM. His first assignment in the new consulting arrangement (he had previously worked for Norman Bel Geddes), was to be this typewriter. It was the beginning of a long and close professional relationship with Watson.

Introduced in 1961, after several years of development, the new typewriter presented a simple, unified form. "In design . . . we tried to emphasize the singleness and simpleness of form by making the whole shape something like that of a stone so that you are aware of the continuity of the sides and under the machine and over the top," Noyes noted.

"Eliot changed our image 180%," said Watson. "He changed our business to something bright and modern. He never gave us a bad piece of advice and he had a way of being terribly right. People would lean on him and often presentations would be opened by saying 'this has been discussed with Eliot Noyes and this is his recommendation.' It was like having something blessed by the Pope."

Master wood model was a still more refined version.

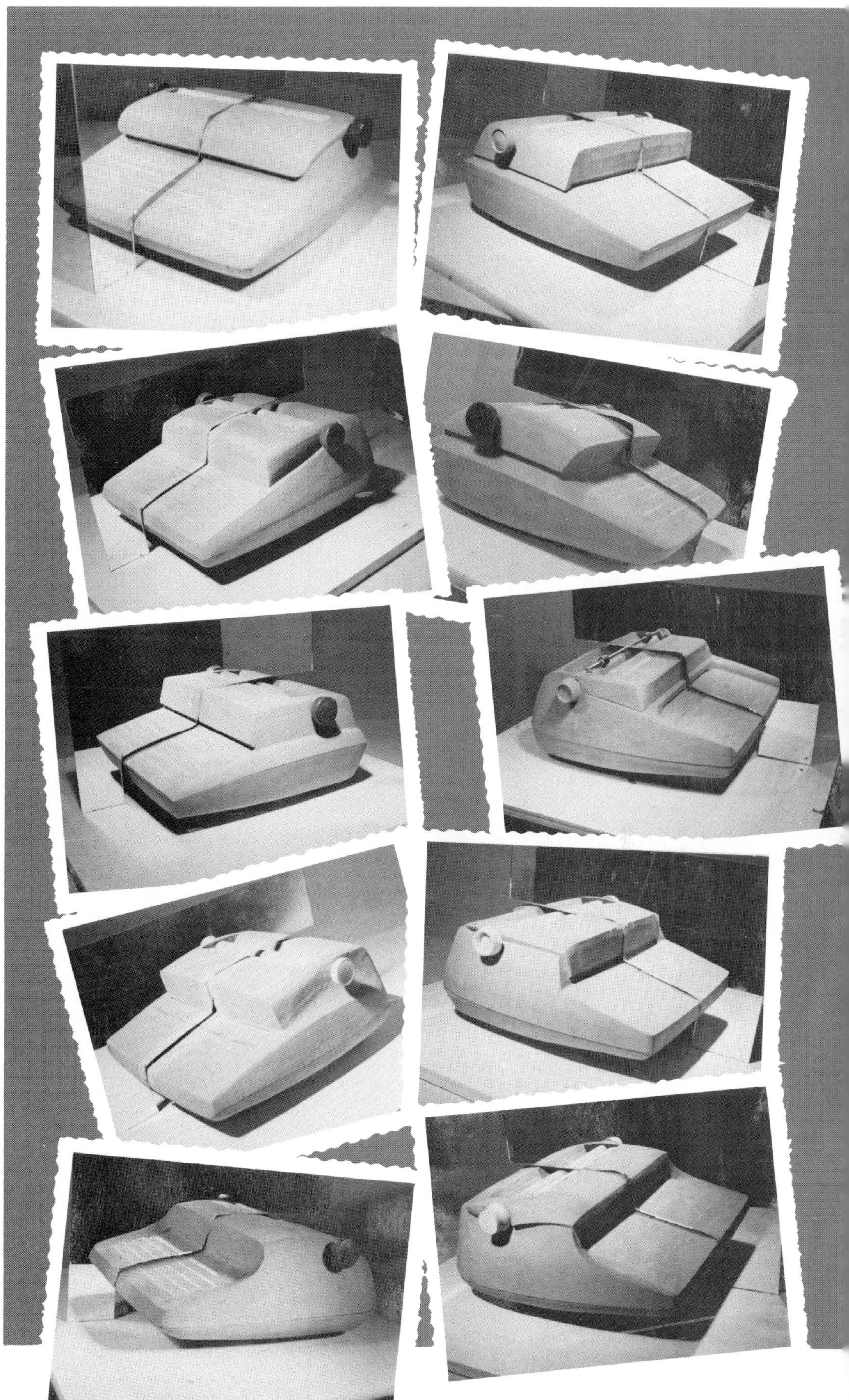

Series of Polaroid photos taken at the Eliot Noyes office as the designers sculpted a basic clay form. Several design directions can be detected, but the bottom photos begin to reveal the concave keyboard and sleek sides that emerged in the final version. Using a mirror, the designers could "work" both sides of the original block and still achieve a complete symetrical look.

Design consultants Eliot Noyes and Associates
Designer Eliot Noyes
Manufacturer International Business Machines Corp.

IBM Selectric I 1961

The first Selectric with "hood" open.

Noyes described sculpting the new machine:

"My work during this phase consisted of making new clay models, wood and plaster models, and finally having sand castings made. This gave us aluminum parts which I then went to work on myself with a hand file to refine the shapes of the top cover and the transition forms between all covers, in view of the fact that this was now an extremely complex molded form and required a great deal of subtlety in its shaping."

Special new keys became the standard for the industry and improved the usefulness of the machine to the typist. Innovations also included a key plate covering the hole between the keys. Color use was new and inventive.

The familiar Selectric type ball.

IBM Selectric III 1980

The typewriter's design was updated in 1980. The Selectric III is an advanced version including new covers, sound reduction attachments, key buttons, controls, knobs, end of page indicator and colors.

It is interesting to note that some IBM executives preferred a competitive design, more like typewriters produced by other companies. It took the perceptive eye of T.J. Watson, Sr., to rule in favor of the Noyes' advanced look.

IBM III (1980). This design is an advanced version including new covers, sound reduction attachments, key buttons, controls, knobs, end of page indicator and colors.

★ GE Toast-R-Oven 1961

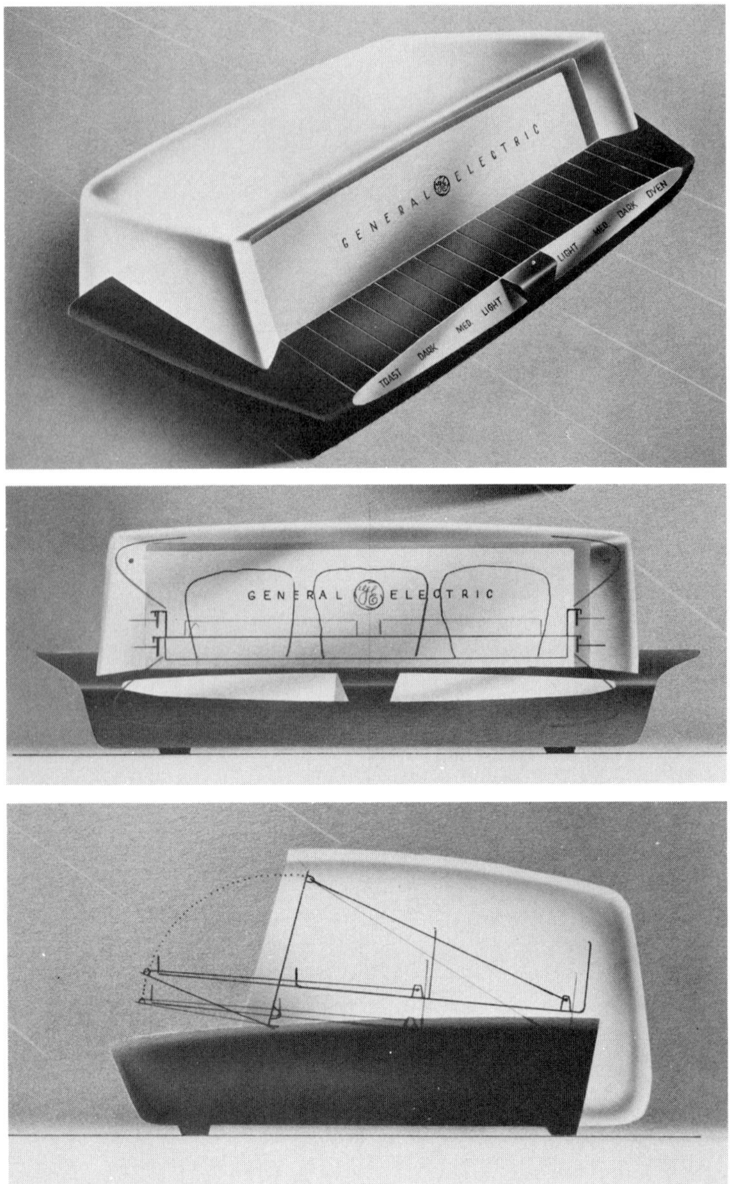

Concept rendering from 1957 based on a three-slice capacity; it was found, however, that units larger than two slices compromised toasting performance

The Design Problem: To develop a two slice toaster that would also heat TV dinners.

The original General Electric design established the type-form for a whole new product category. The Toast-R-Oven has grown in size and the controls have become more sophisticated, but the basic concept has remained the same over the years.

Based on a two-slice concept, an engineering "bread board" model was built in 1958.

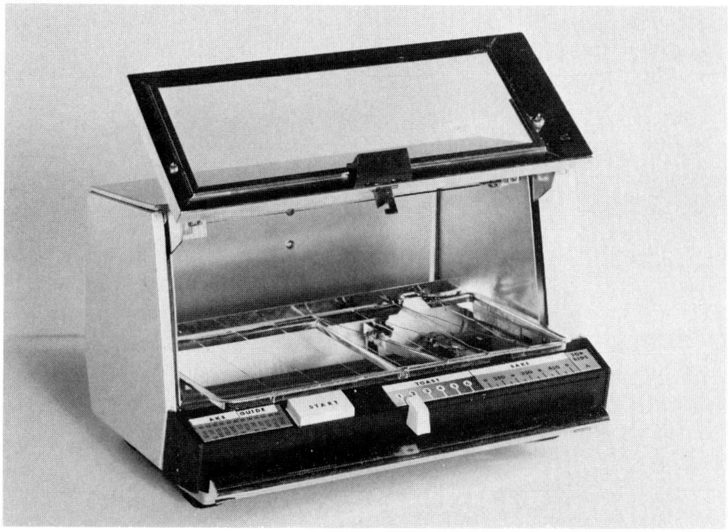

The 1972 redesign.

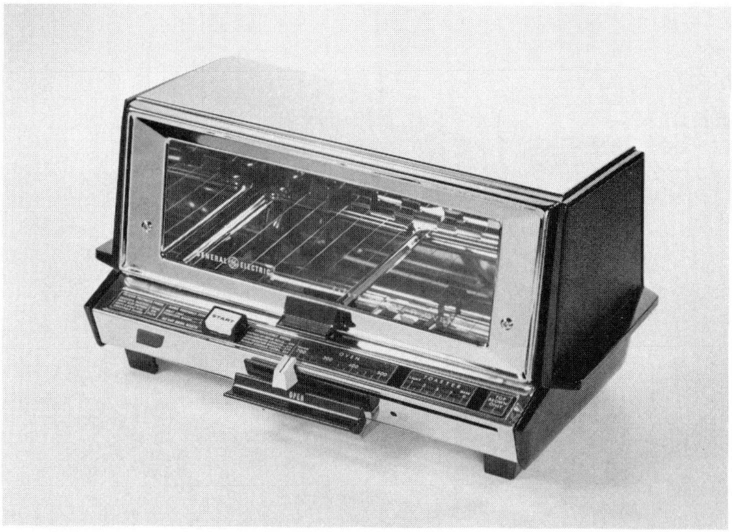

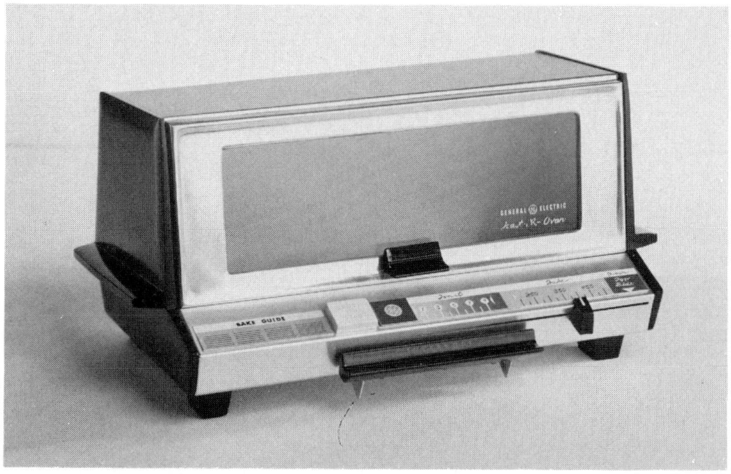

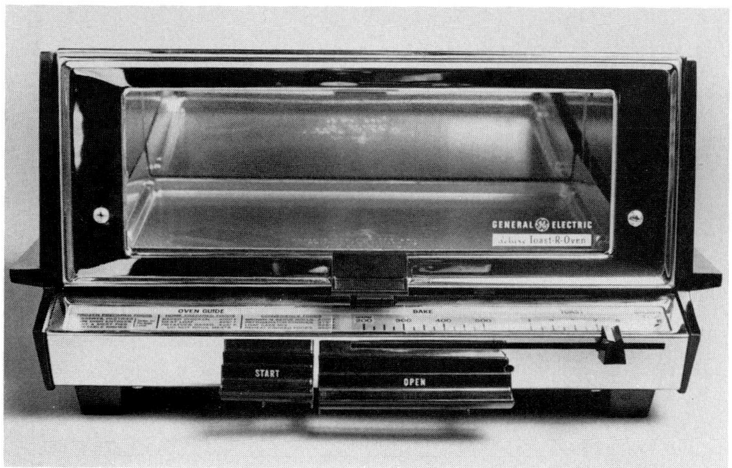

Manufacturer General Electric Company
Designers Paul O. Rawson, Roger L. Funk
Engineer William Schmall

The original T-93 Toast-R-Oven was introduced in 1961.

The current 1983 model, twenty-two years later, has easier-to-read information, better switches and incorporates a warming tray on the top.

Kodak 550 Carousel Projector 1961

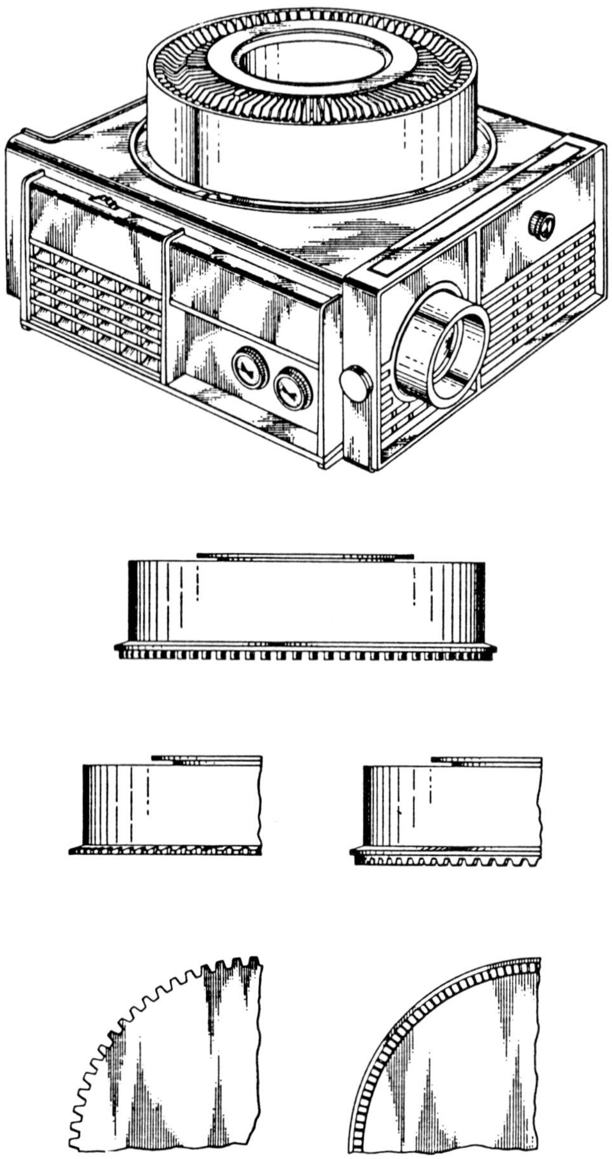

Patent drawings from 1963 for a combined slide projector and slide tray by Richard J. Olson and Frank A. Zagara assigned in 1961 to Eastman Kodak.

The Design Problem: To develop a system to handle slides reliably that would provide accessibility for editing use gravity for slide handling and provide a way of dropping slides out of the bottom of the slide tray, with an elevator for re-insertion.

The major decision was whether the slide tray should be circular or straight. The answer would define the product. Circular trays provided sequential projection, accuracy of slide location from a central point and rapid indexing to any slide. A disadvantage was the larger volume circular trays occupied in comparison to straight units. The industrial design department established the basic format of the circular tray and sold the idea to management through drawings and models. Coordinated efforts with engineering brought about successful production.

The original 550 carousel projector of 1961.

Manufacturer Eastman Kodak Company
Industrial design department head Arthur Crapsey
Industrial designers Richard Olson, Frank Zagara
Project engineer Herbert Robinson
Drafter/designer Matthew Dipetro

Kodak 550 Carousel Projector 1961

The introduction of the first Carousel in the early 60s reduced the difficulties previously associated with slide projection. The circular tray held a larger number of slides than other available models, the editing process was simplified and the rate of projection increased in speed.

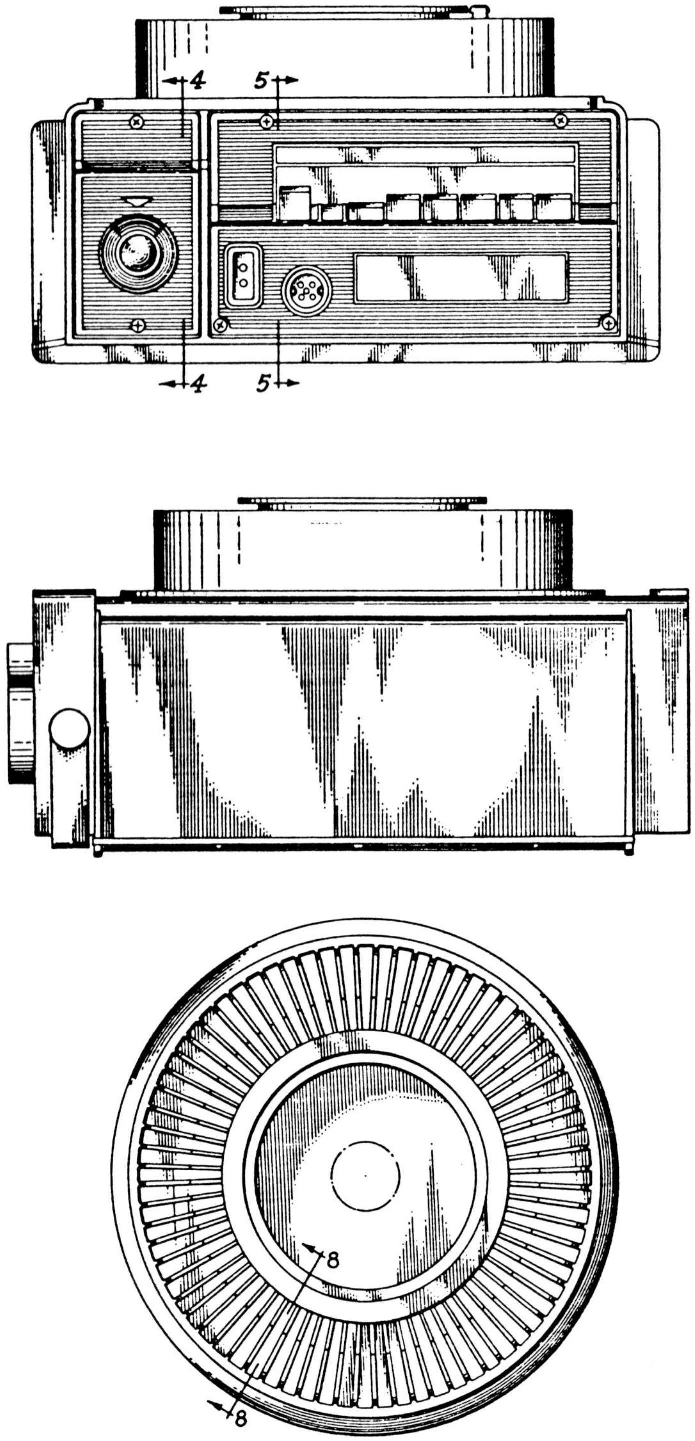

Patent drawings from 1963 for a combined slide projector and slide tray by Richard J. Olson and Frank A. Zagara assigned in 1961 to Eastman Kodak.

An array of early concepts for the carousel projector.

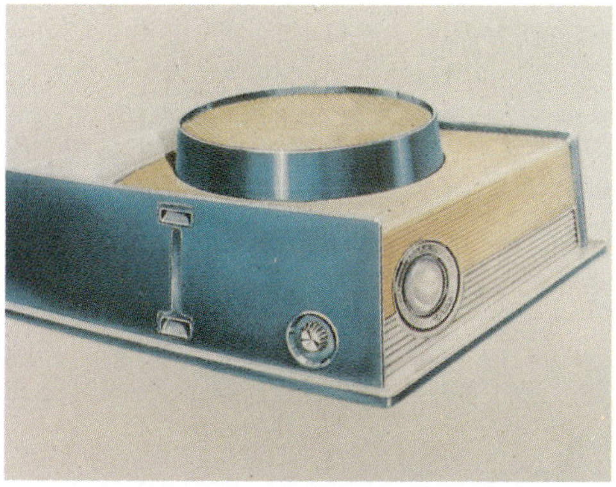

The 855 carousel projector of 1975.

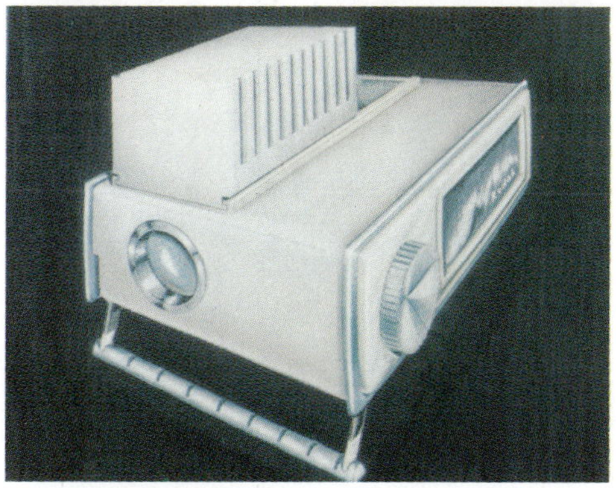

The 5600 carousel projector of 1980.

Kodak Ektagraphic III Projector 1981

The 860 carousel projector of 1969.

The circular tray format initiated a 23-year series of redesigned projectors that have improved, but not effectively changed the original function. The 1983 EKTAGRAPHIC III design is structured to accommodate the slide handling needs of the photographic beginner and the specifications of the audio-visual professional. Consolidation and movement of controls to the side of the projector make sense from a human factors point of view and establish a format that is distinctly different from Carousel projectors of the past. The carry handle is also incorporated into the design.

Manufacturer Eastman Kodak Company
Industrial design department head Richard J. Olson
Industrial designer Richard Weiss

Stanley Powerlock Power Return Rule 1963

The Design Problem: To design a high quality power return rule with locking feature suitable for the professional and home market.

In 1962, the Powerlock constituted a definitive break with tradition. Previously, the typical push-pull rule was contained with a D-shaped case. It was not self-winding, nor did it have a locking device. The new spring-loaded locking rule, the invention of Robert West at Stanley Tools, presented a challenge to the industrial designer.

The solution, which follows various preliminary studies, comprises a shape that fit the human hand, with easy access to a thumb lock (shaped for two-directional action) and smooth, highly polished sides with textured edges for gripping. The case is recessed for easy and accurate reading of the rule. Cases vary in size, the smallest containing 1¼-inch by three feet rule blade, the largest, one-inch by 30 feet.

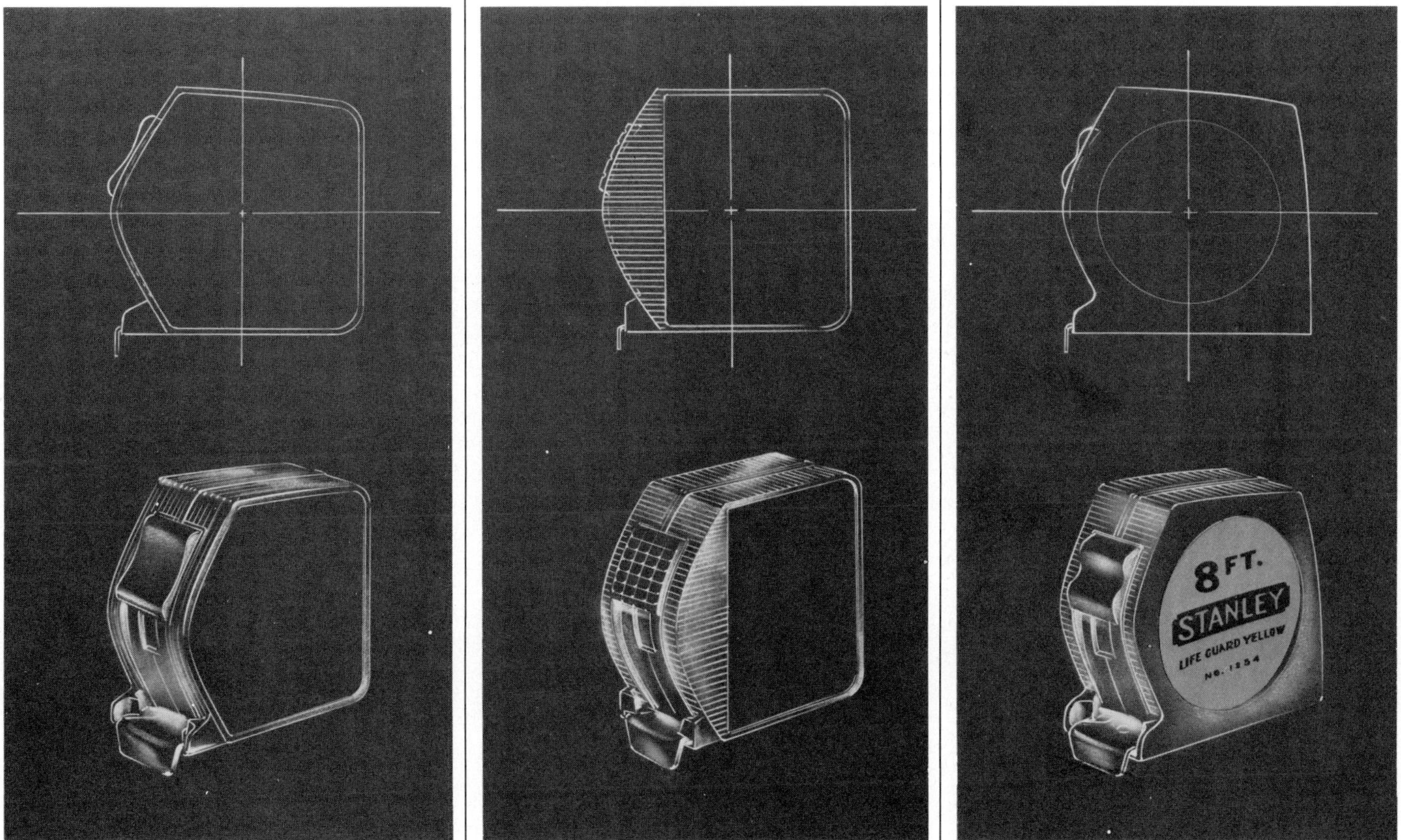

Renderings of three alternative designs.

Manufacturer Stanley Tools, Div. of The Stanley Works
Design consultant Laird Covey
Manager of new product research and engineering Robert F. West
Product line engineering manager Edward C. Rutty
Design engineer Frank Czerwinski

The shape fits the thumb and hand, with easy access to the thumb lock and textured edges for gripping.

Polaroid Land SX-70 Camera 1972

The Design Problem: To develop an instant camera/film system that would allow users to concentrate on the picture rather than on the process of making it and to achieve a compact and elegant design.

Dr. Edwin Land's determination to achieve these goals pushed the state of the art technology to its limits. Polaroid's extensive research in chemistry, optics, electronics, materials and processes had resulted in breakthroughs that made the camera possible.

The Dreyfuss office was part of the design team from the beginning. The challenge was to keep the industrial design input abreast of research and development (often the camera's configuration was changed on a daily basis). The Dreyfuss staff worked with Polaroid scientists and engineers to incorporate the new technology into camera design, while maintaining design integrity.

Two decisions by Dr. Land reduced the overall size of the camera: building the power source into the film pack, which eliminated the need for a battery compartment; incorporating a through-the-lens viewing system (single lens reflex), which enabled most of the view-finder to be internal. (The latter required a small 'penthouse,' which was then designed as a grip to open the camera). Materials research led to the use of chrome-plated plastic for the body shells. The plating provided the required stiffening for the plastic parts. Weight and cost were reduced without compromising.

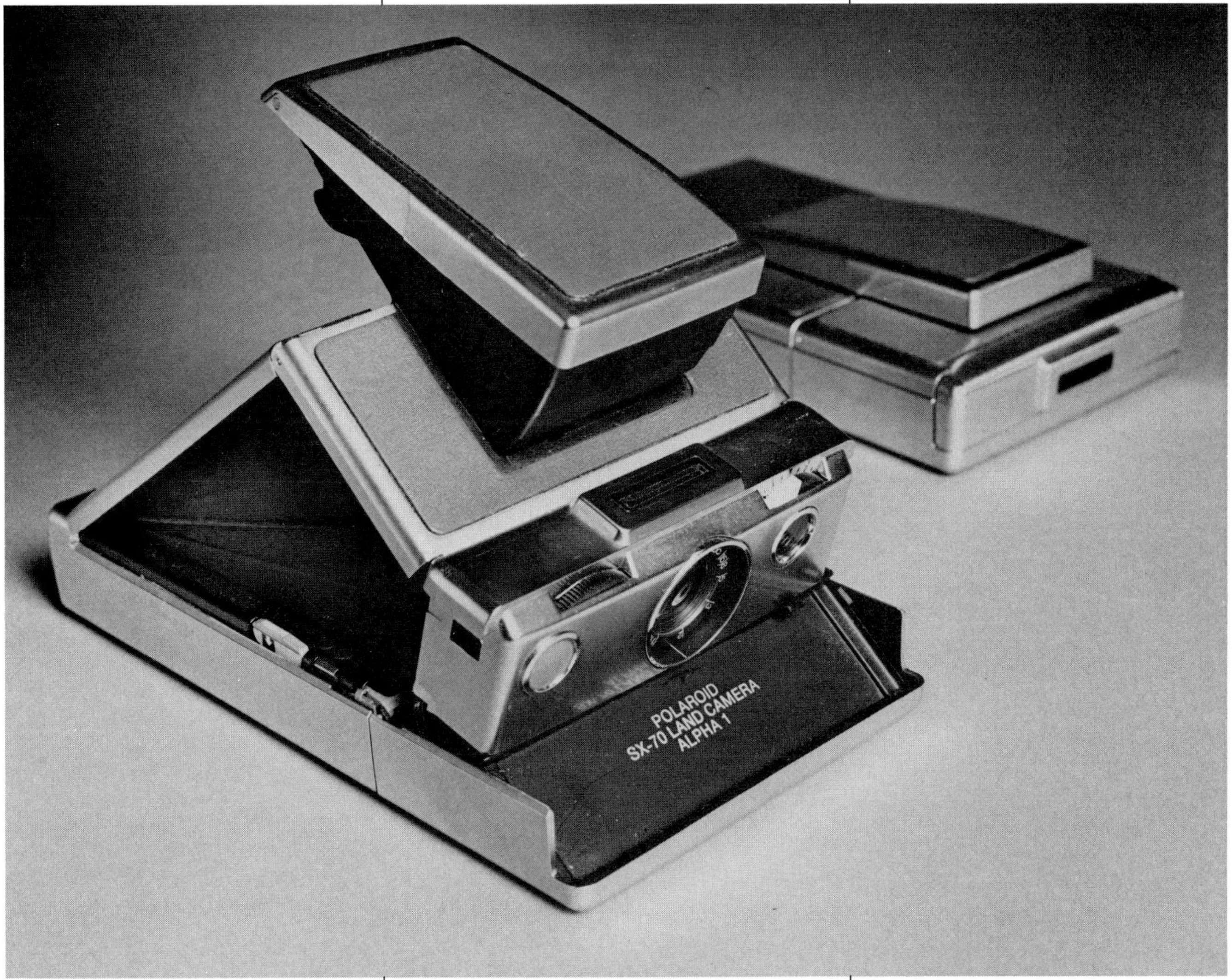

The Land SX-70 introduced a whole new era of photography and took the utmost cooperation between the inventor, Dr. Edwin Land, and the Dreyfuss design team.

Design consultants Henry Dreyfuss Associates
Advisor Henry Dreyfuss
Designer (partner) James Conner
Manufacturer Polaroid Corp.
Chairman, president and inventor Dr. Edwin H. Land
Support Research, engineering, production and marketing staff at Polaroid

Instrumentation Laboratory Statsep Plasma Separator 1981

First design of interior mechanism.

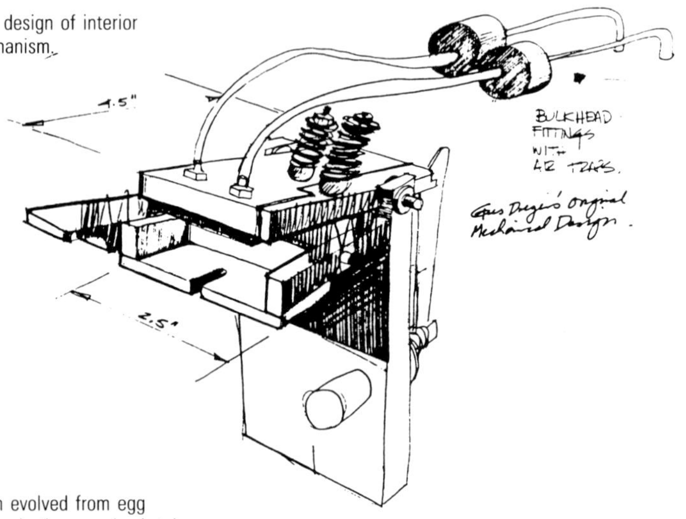

Form evolved from egg shape in these early sketches.

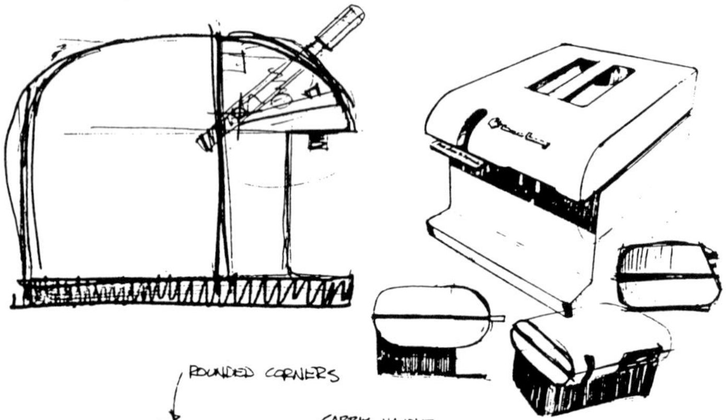

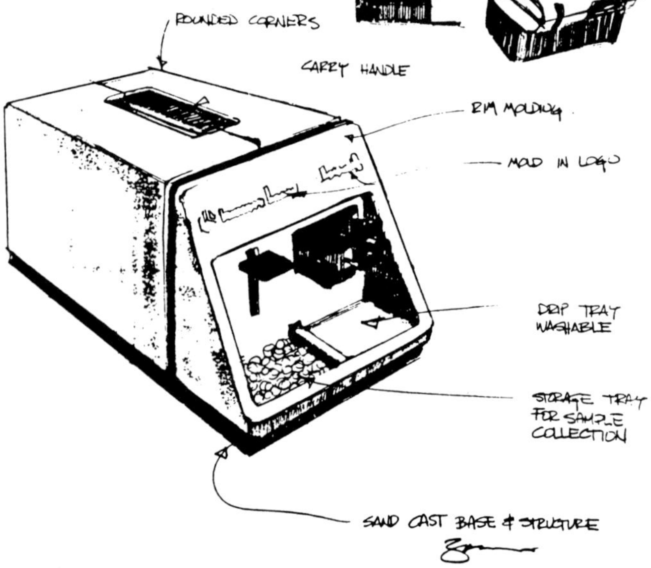

The Design Problem: To achieve acceptance by hospital staffs for a new medical product.

Instrumentation Laboratory had developed a new technology to separate plasma from whole blood in less than 90 seconds. Although there was a definite need for a system that operated quickly and at low cost for emergency diagnostic testing in hospitals, the company was uncertain about the market and user acceptance of their technology. Through on-site testing and observation of early engineering models, designers determined that the use and maintenance of the device had to be simplified, so that it would be less threatening to the busy nurses and technicians who would use it.

Data gathered from users established that simplicity, clarity and friendliness were as important as accuracy and technological sophistication. In the redesign, all unnecessary components were eliminated, each part was designed to perform as many functions as possible, and the points of contact between user and machine were refined. The designer, Gianfranco D. Zaccai, notes that once all the requirements were identified, the appropriate form was considered and "the image of the egg presented itself. It is strong and structural; it is simple in form; houses a great deal in a small volume and looks friendly. I decided to take this form, square it up slightly, cut away the front to create a conveniently inclined control panel and suspend the internal components within the 'shell'."

Zaccai feels that the needs and wishes of a diverse audience—manufacturer, seller, buyer, user and repairer—should be addressed within the limitations of technology. On the other hand, he continues, "minimally addressing all these concerns is not enough. Once the product is made and used, it touches the user, for better or worse, on an emotional level. The art of the design process is to make this interaction a positive experience."

The designer's task was to simplify a complex medical instrument.

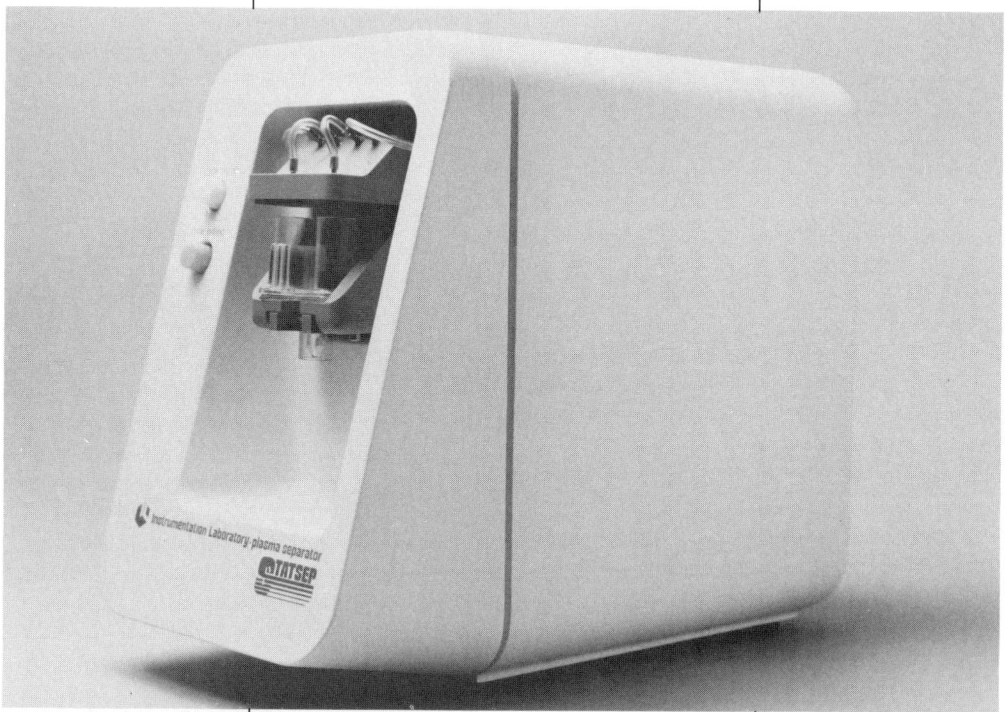

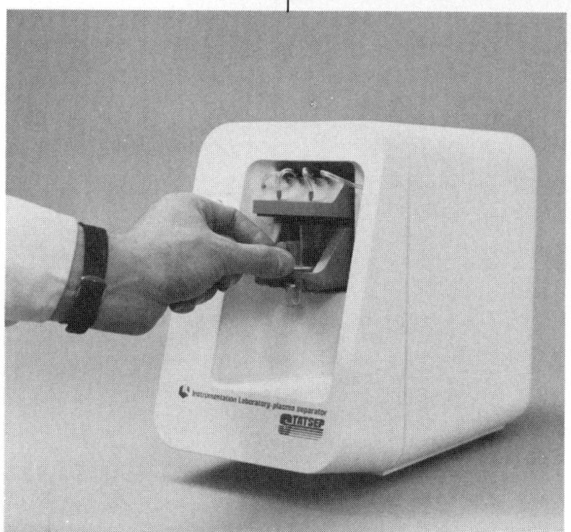

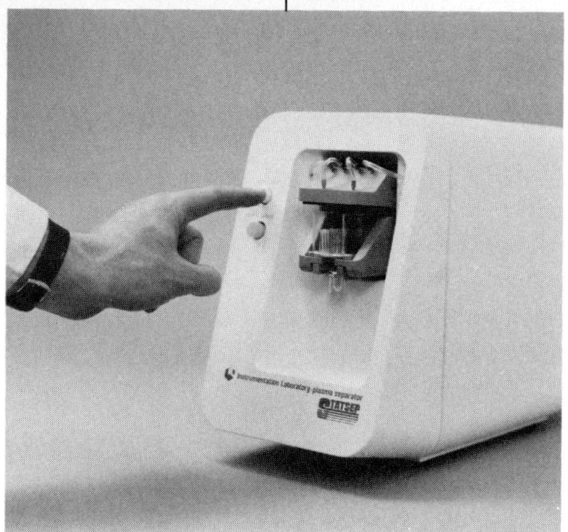

Manufacturer Instrumentation Laboratories, Inc.
Manager of industrial design Gianfranco D. Zaccai
Industrial designer/model maker Richard Holtz
Staff scientist Kenneth Legg PhD
Mechanical engineers Gustav Drier, David Wheinstone
Mechanical designers William Flynn, Leo Patrician
Product managers Richard Pinkowitz, David Williams

★ Litton Pocket Socket Socket Wrench 1979

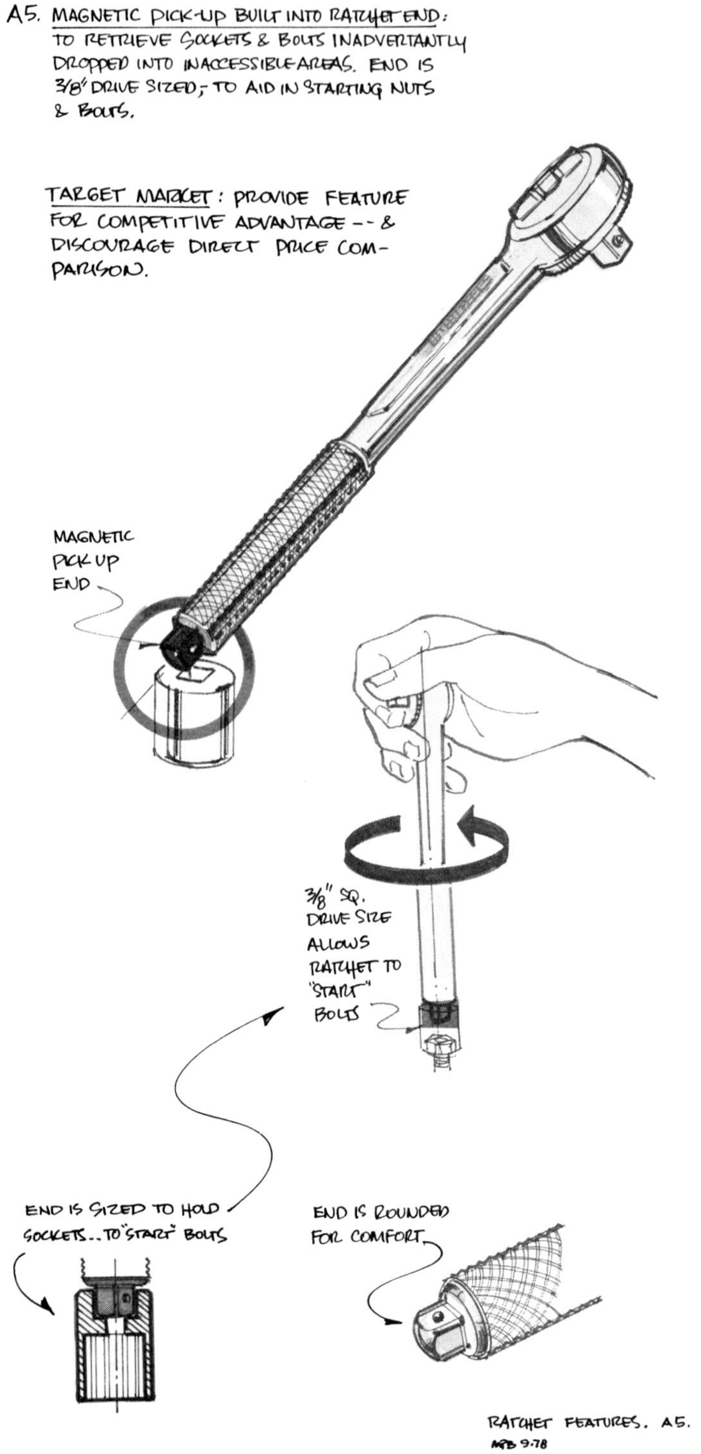

The Design Problem: To invent an improved socket wrench for use by consumers, auto mechanics and machinists.

Innovations & Development was responsible for the entire design process from concept through design and engineering. Market research and observation revealed that many of the problems incurred while using socket wrenches would be eliminated if the sockets were held within the ratchet. During brainstorming sessions, the idea of putting sockets into the handle of the ratchet was generated. The most important consideration was a functionally improved ratchet. Other design criteria included cost, human factors (the handle had to be small enough to be workable while containing 5 sockets) and appearance. Various engineering concepts were considered and rejected before reaching the final solution, which was to insert-mold a stamped piece of steel into an ABS housing.

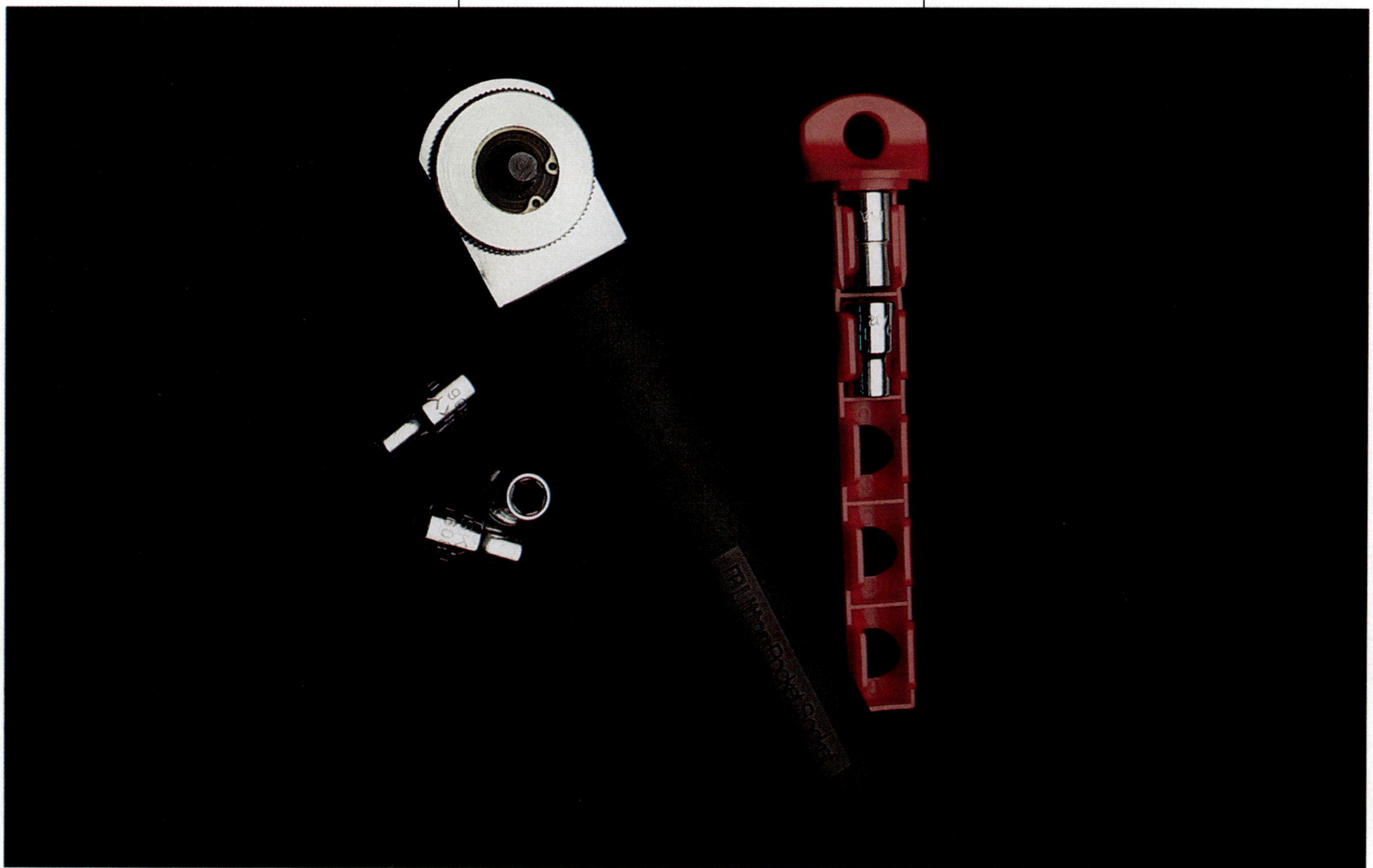

In the final version, sockets are stored snuggly in a plastic frame that withdraws from a round handle.

Design consultants Innovations & Development, Inc.
President and project director Edward Levy
Vice presidents Edward Meisner, Gary Grossman
Design director Michael Ballone
Manufacturer Litton Industries, Inc.
Vice president of marketing—Litton John Trebel

Magnavox Compact Disc Player 1983

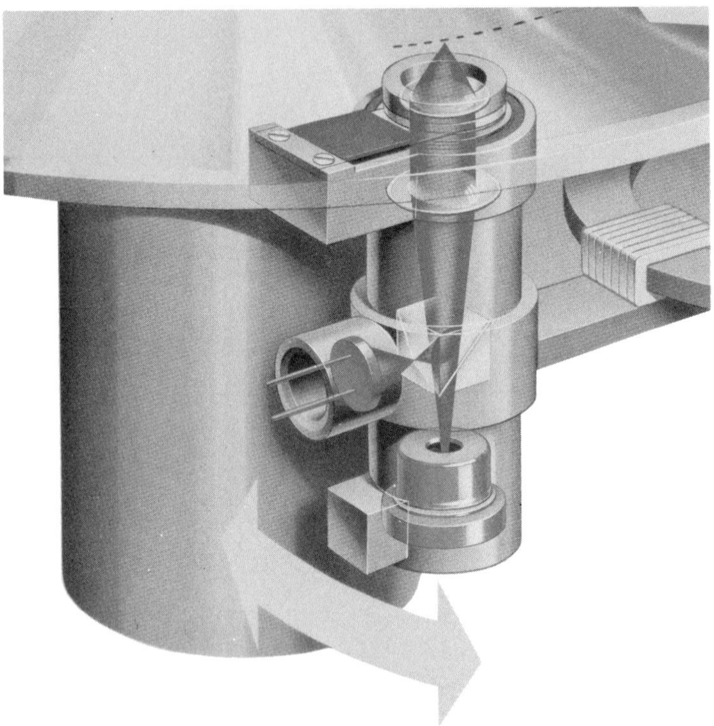

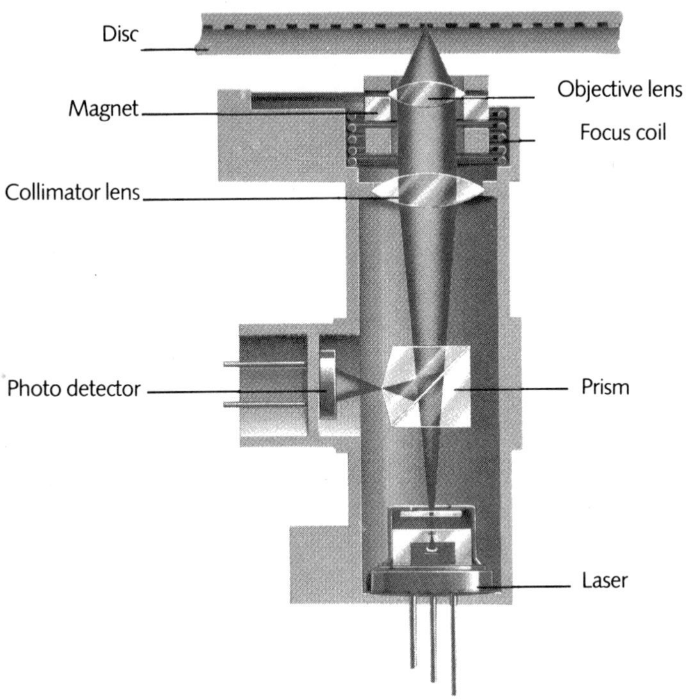

The Design Problem: To develop an easy-to-use audio player for the new compact discs.

In the late 60s, research began into a new type of recording device that resulted in the development of the compact disc. This was the first truly technological breakthrough in sound reproduction since Thomas Edison invented the phonograph in 1877.

With the compact disc, there is no physical link between the player and the actual recording. The original sound is digitally encoded in the form of microscopic pits and flat areas along a 2½ mile spiral track on the disc, then sealed with transparent plastic that protects against dust, dirt and scratches. The player uses a low-power, solid state laser scanner to "read" the recording and produce the ultimate in pure, undistorted sound.

This disc player was designed to accommodate the new compact discs. Meant to relieve the high-tech look of previous units, it is designed for user convenience and ease of operation. The unit has a motorized front loading drawer designed especially for use in rack audio systems; a digital elapsed time display, and quick track access, multimode programming. It is compatible with any hi-fi system.

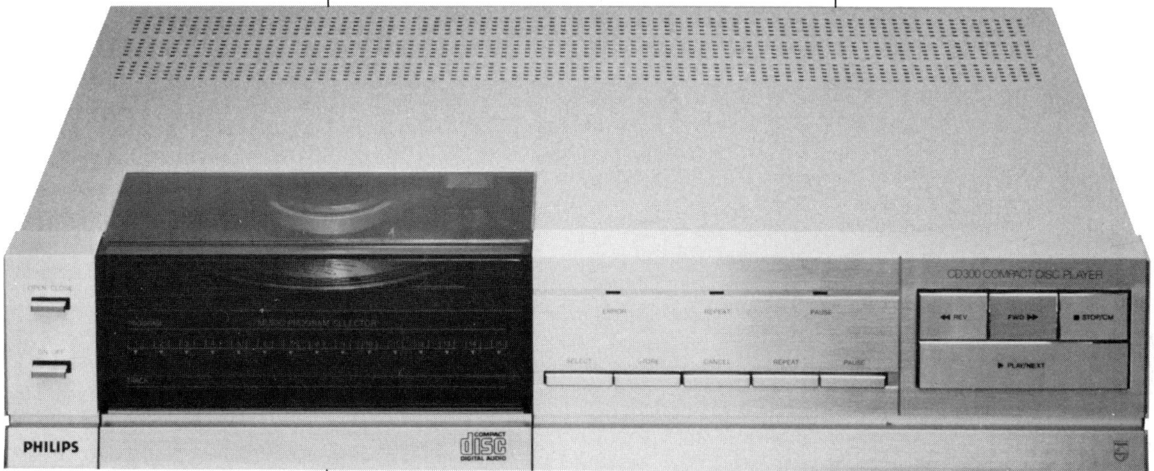

Meant to relieve a high-tech look of previous units, product is designed for user convenience and ease of operation with a motorized front loading drawer.

Manufacturer N.V. Philips of The Netherlands
Design director Robert Blaich
Support Philios design staff

AMERICAN

2

FUNCTION FORM

DESIGN

CLASSICS

GF 40/4 Stacking Chair 1964

The Design Problem: To design a functionally more useful dining or side chair than had ever existed.

Working independently, Rowland spent eight years developing this chair. Many full size prototypes were built during the process; the early prototypes were not successful since manufacturers would not risk re-tooling for such a new concept.

The pivotal moment occurred when Rowland discovered that a stackable chair with thin legs would be feasible. Sitting on a single chair and finding it reliably rigid, he then stacked it on an exact duplicate and calculated that, in quantities, two chairs would store in less than one cubic foot. (40/4 refers to the fact that forty chairs can be stacked within four cubic feet).

Aluminum and plastic seats and backs were explored and rejected in favor of formed steel and molded wood veneer.

The chair concept became a reality with the input of GF engineers, tool designers and an enthusiastic managment. The 40/4 chair is in the Design Collection of the Museum of Modern Art.

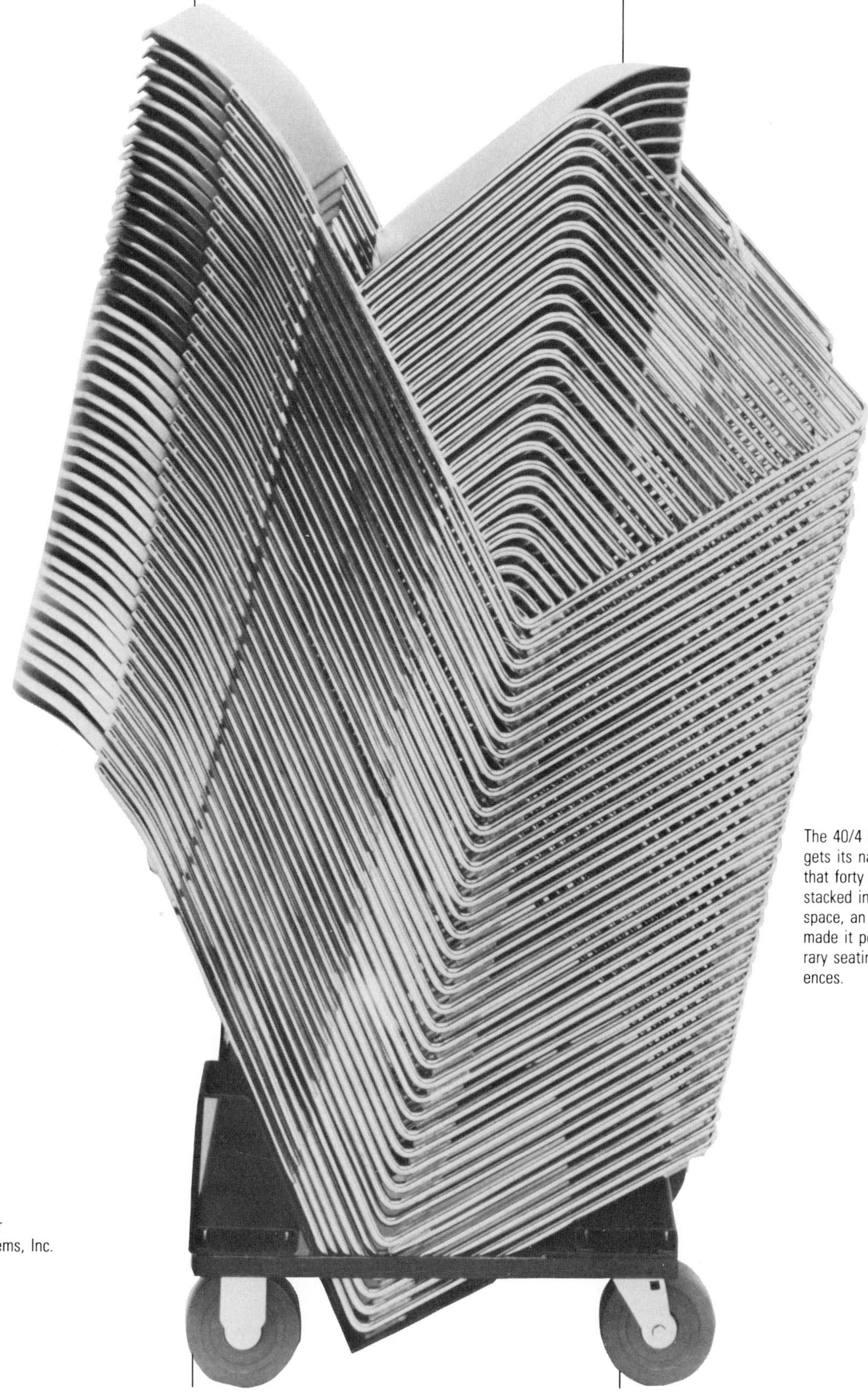

The 40/4 Stacking Chair gets its name from the fact that forty chairs can be stacked in four cubic feet of space, an attribute that has made it popular for temporary seating for large audiences.

Designer David Rowland
Technical director Rudy Wilmer
Manufacturer GF Furniture Systems, Inc.

AT&T Trimline Telephone
1965

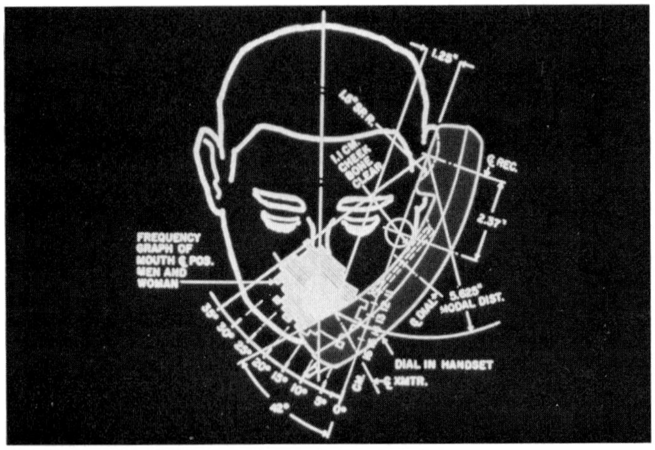

Throughout the process, human factors requirements, developed in the Dreyfuss office, are considered.

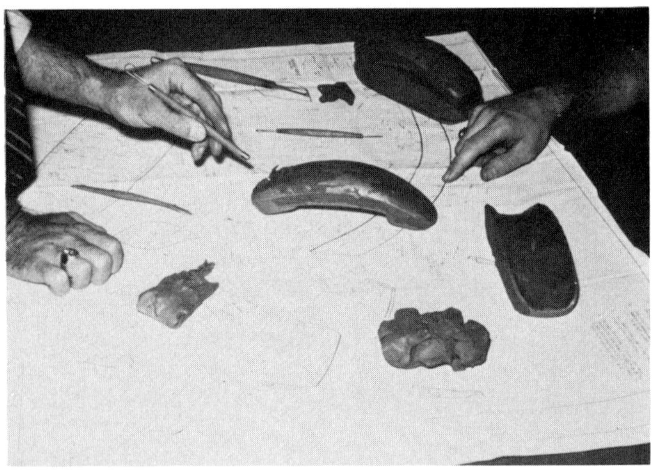

Clay models aid in establishing the form.

The Design Problem: To design a compact, dial-in-handset telephone for residential use, in both wall and table versions, with standard rotary dial or "touch tone" push button dial. The design should fit various decors and meet appropriate human factors standards for use.

Work on the Trimline development began in 1959 when AT&T surveys indicated a potential market for this product. Existing phones were studied world-wide, as well as user feedback from a "lineman's" set (a type of dial-in-handset used by Bell System operating company personnel). Human factors research was based on anthropomorphic data assembled over the years by the Dreyfuss office and Bell Labs, as well as user tests. Weight was very important; the handset had to be light and the base weighted for stability. Research indicated that the user preferred a "base" to return the handset to, rather than a one piece set, which was mechanically feasible.

The handset had a critical size that could not be exceeded. The rotary dial version with its finger holes of a given dimension and webs in between was too large in diameter and forced the handset into unmanageable proportions. The idea of having the finger stop *move* as the dial was cranked eliminated the space between #1 and the finger stop, thereby reducing the diameter and saving the project.

Don Genaro points out that "for the industrial designer, the distance between the drawing board and the assembly line is not one inspired leap, but rather a series of careful and patient steps with close attention paid to every detail along the way."

AT&T Genesis Telesystem 1983
The most recent Dreyfuss and Bell Laboratories collaboration has produced the "Genesis," a total communication and information management system incorporating the latest technology. Don Genaro cites the growing importance of "user friendliness" in design. Today it is assumed that the product must "feel good." What is needed, he believes, is a growing understanding of the psychological aspects of the consumer's confrontation with the new technologies.

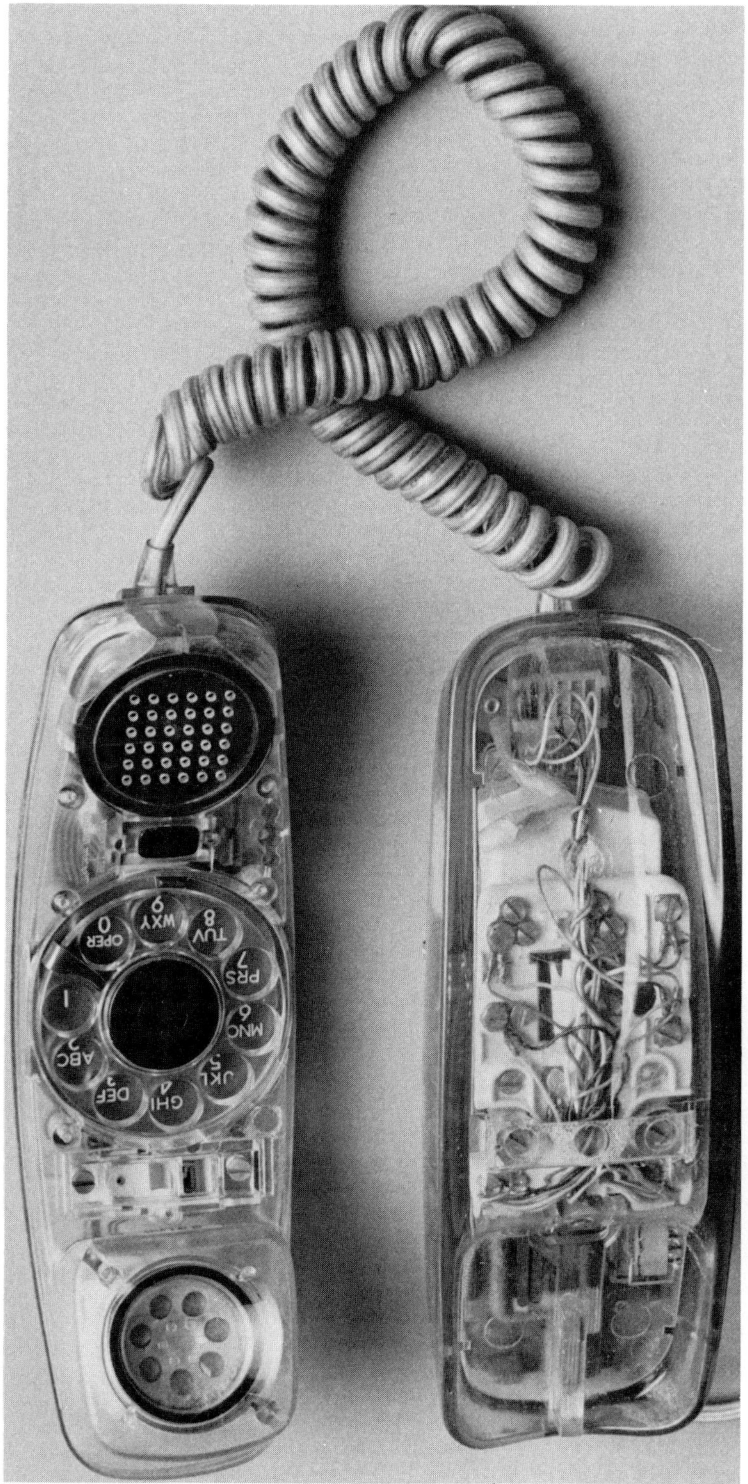

Three dimensional studies are made for component placement.

Design consultants Henry Dreyfuss Associates
Designers Donald M. Genara, Alvin R. Tilley
Bell Laboratories Robert E. Prescott, Leslie Wilder
Manufacturer Western Electric Company (AT&T)

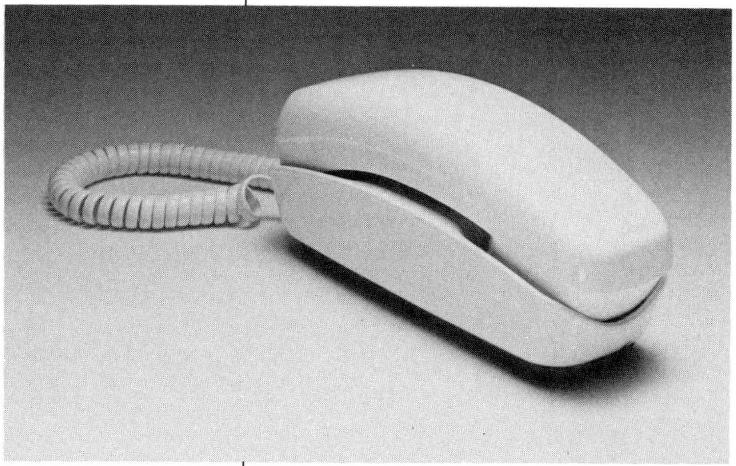

Development of the Trimline began in 1959 and was a radical departure from the classic 1937 and 1951 models, both also design by the Dreyfuss office.

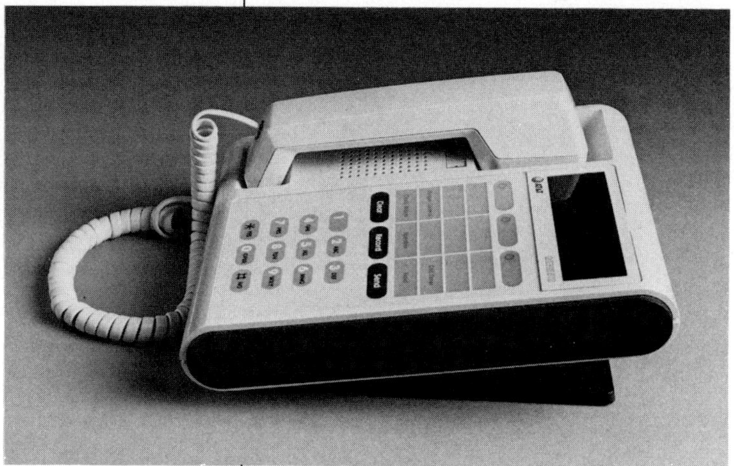

AT&T's Genesis Telesystem combines communications and information management using the latest technology in a "user-friendly" design.

Mobil Gas Pump 1966

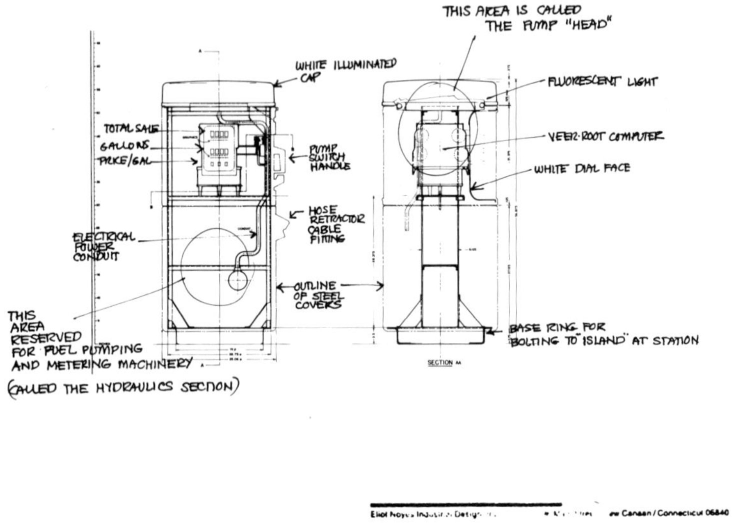

Early development drawings from the Eliot Noyes office.

The Design Problem: To create a more attractive and efficient gasoline pump as part of an effort to clarify overall service station design.

Eliot Noyes was commissioned to prepare a comprehensive design program for Mobil's world-wide operations that would combine graphic and industrial design with architecture. The first thing he discovered when he studied the ugly, cluttered Mobil stations in 1964 was that there was often so much visual distraction for the driver as he turned onto the station apron that he could only find his way to the pumps by following arrows on the ground.

The pump and the circular canopy were designed by the Noyes office. The basic Mobil logotype, with its now famous red letter O, was designed by Chermayeff & Geismar.

The new Mobil pump was the solution for part of a complete redesign of the service station.

Design consultants Eliot Noyes & Associates
Designers Eliot Noyes, Ernest Bevilacqua, Robert Graf, Allen Hawthorne
Client Mobil Oil Corp.
Mobil corporate design and graphics George Priggen
Mobil engineering Mott Prudames

Edmund Scientific Astroscan Telescope 1975

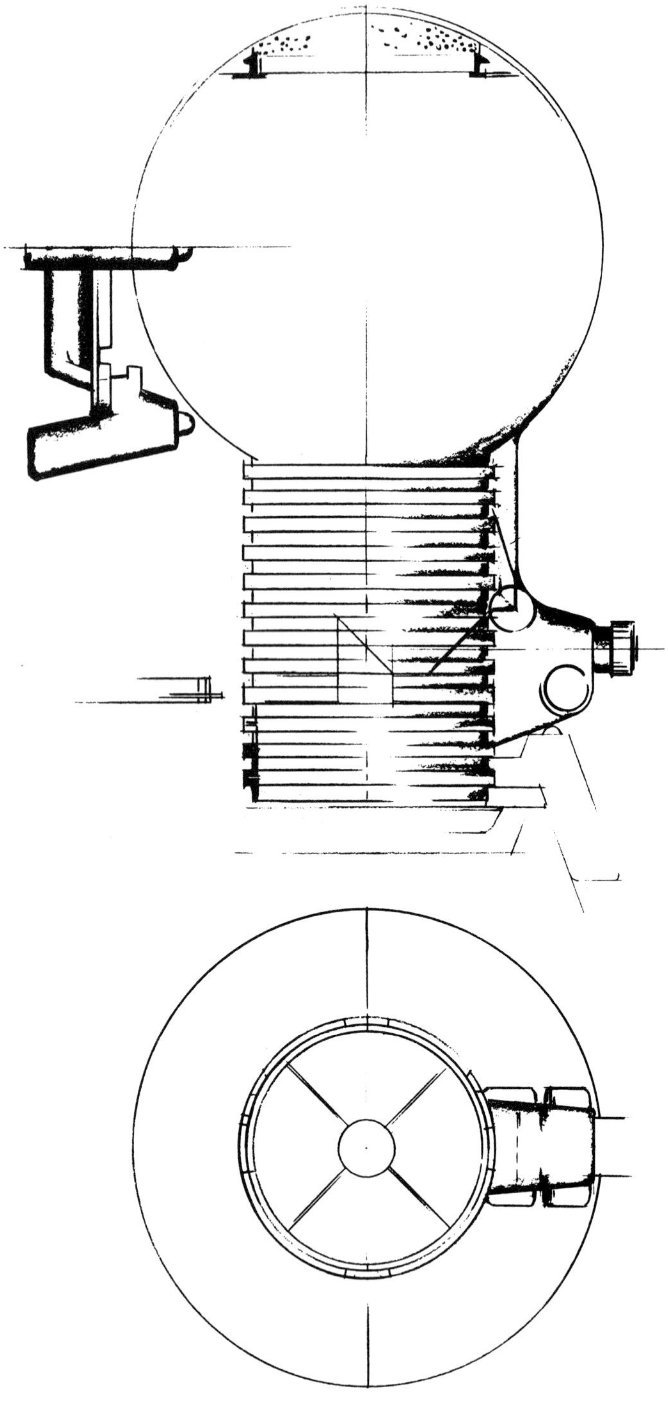

Preliminary engineering drawings.

The Design Problem: To design a mass produced, high quality, wide angle telescope that would be low cost, portable and satisfying to a wide range of consumers.

Early research with the client and astronomical consultants clearly established user and functional priorities.

The turning point in the project came in three stages. First, the decision to shape the two halves in a common mold with an insert for the viewfinder (engineering requirements specified that the eye piece be part of the body). Second, the conclusion that a clean, business-like, professional design would result in greater respect for the product (this was suggested by astronomical research). Third, the selection of red, the color which would set a bold corporate image, and also stand out well in the retail store. Red also tends to absorb light during operation at low light levels (red is perceived as black, thereby not altering the night vision of the user).

This is an excellent example of an integrated design solution. Industrial design developed the manufacturing process and determined materials, assembly process, human factors dimensions and appearance; engineering contributed optical design and theoretical engineering support; astronomical consultants contributed optical and market criteria.

Red creates a strong image, but is perceived as black at low light levels so that the unit would not alter the user's night vision. Engineering required the eyepiece to be part of the body, which led to this robust design.

Design consultants Peter Bressler Design Associates
President Peter Bressler
Manufacturer Edmund Scientific Company
President Robert Edmund
Director of engineering Kevin Noran
Staff engineer Michael Simmons

★ Welch Allyn Otoscope and Opthalmoscope 1966

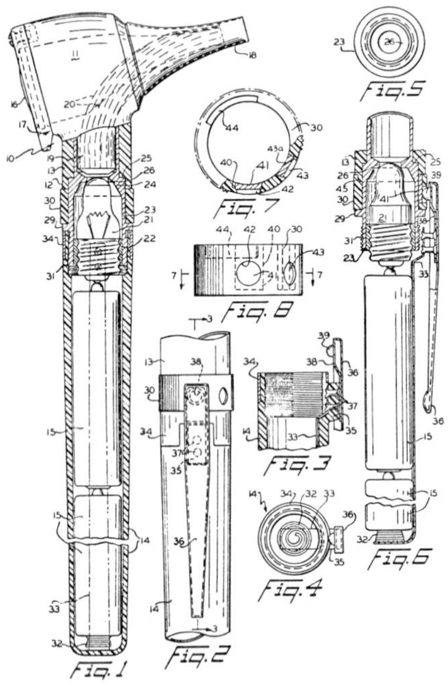

Patent drawings reveal use of fiber optics in these diagnostic instruments.

The Design Problem: To reconceive diagnostic instruments using fiber optics and penlight batteries to reduce the product from traditional flashlight size and technology to fountain pen scale.

In redesigning the Opthalmoscope (an occular diagnostic instrument) and the Otoscope (an oral diagnostic instrument), client and designer agreed not to upgrade the appearance of older instruments, but to attract new markets for a penlight size unit that could be carried easily by doctors. The use of fiber optics in the Otoscope put the light bulb in the handle instead of in its original location in the middle of the doctor's line of vision.

It was important that the product have a technical rather than commercial appearance to offset resistance of doctors and patients to the new configuration of a traditional instrument. The form, though smaller, had to convey the dignity and dependability of the original.

The instruments were selected for the Design Collection of the Museum of Modern Art.

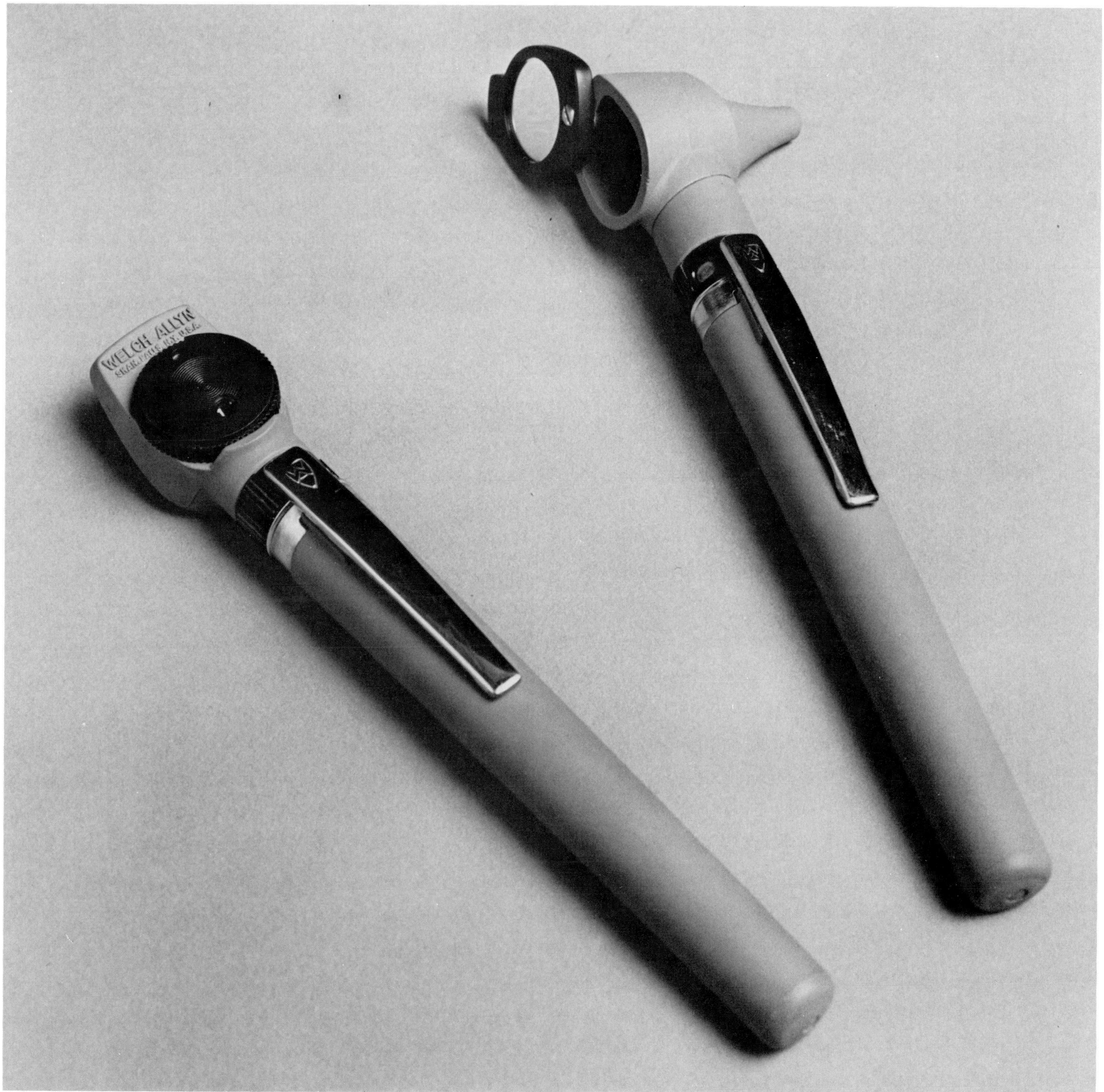

Design consultants Pulos Design Associates
President Arthur Pulos
Senior associate Paul B. Sweeney
Associate Richard Chapman
Manufacturer Welch Allyn, Inc.
Director of engineering William Moore
Project engineer Richard Newman
Engineer John Conners

Designs for the Otoscope (left) and Opthalmoscope (right) encase new technology in form familiar to doctors.

Xerox Primary School Microscope 1971

Detail from engineering drawing.

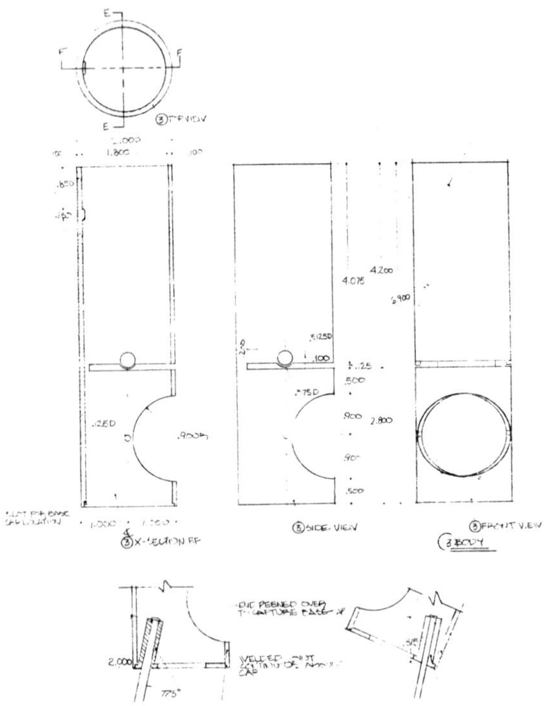

The Design Problem: To design a microscope to be included in a science kit for primary school children that could be produced at minimal cost, be virtually indestructible, have a magnification power of 100x and be self-instructive.

In setting forth the problem, Xerox initially considered a product that would imitate a laboratory microscope. In the course of development, the client accepted the designer's suggestion that the microscope take the most appropriate shape for meeting the needs of the assignment. The result is a simple, elegant microscope in which the visual and physical orientation makes the mode of operation self-evident.

In an innovative design feature, the ring on the base which holds the microscope is set into the cylinder at an angle so that when it is rotated at 180°, it changes from a tilted position suitable for a dry slide, to a perfectly vertical position suitable for a wet slide. This effective and sturdy microscope was produced for $1.50.

In discussing his approach to design, Gersin explains, "Ultimately . . . we are designing for our client's audience, the user. In order to do this successfully we learn as much as we can about the user's needs, desires and perceptions . . . it is often the information we acquire from the end user, through research, that we find most beneficial."

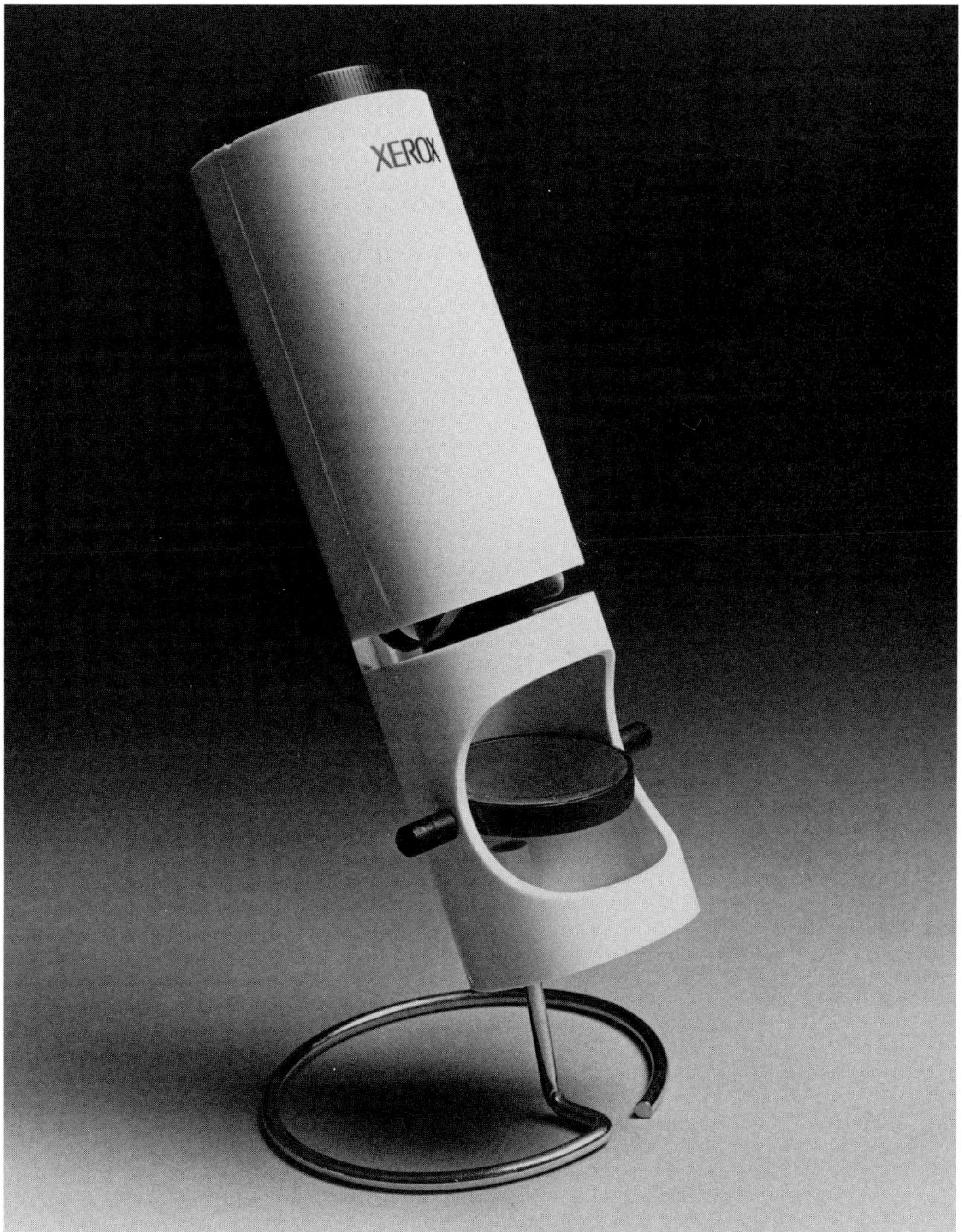

The elegant ring base, in-set at an angle, rotates 180 degrees to accomodate dry or wet slides.

Design consultants Robert P. Gersin Associates, Inc.
Design director Robert P. Gersin
Manufacturer Xerox Corp., Learning Systems Division

★ **Acoustiguide** 1974

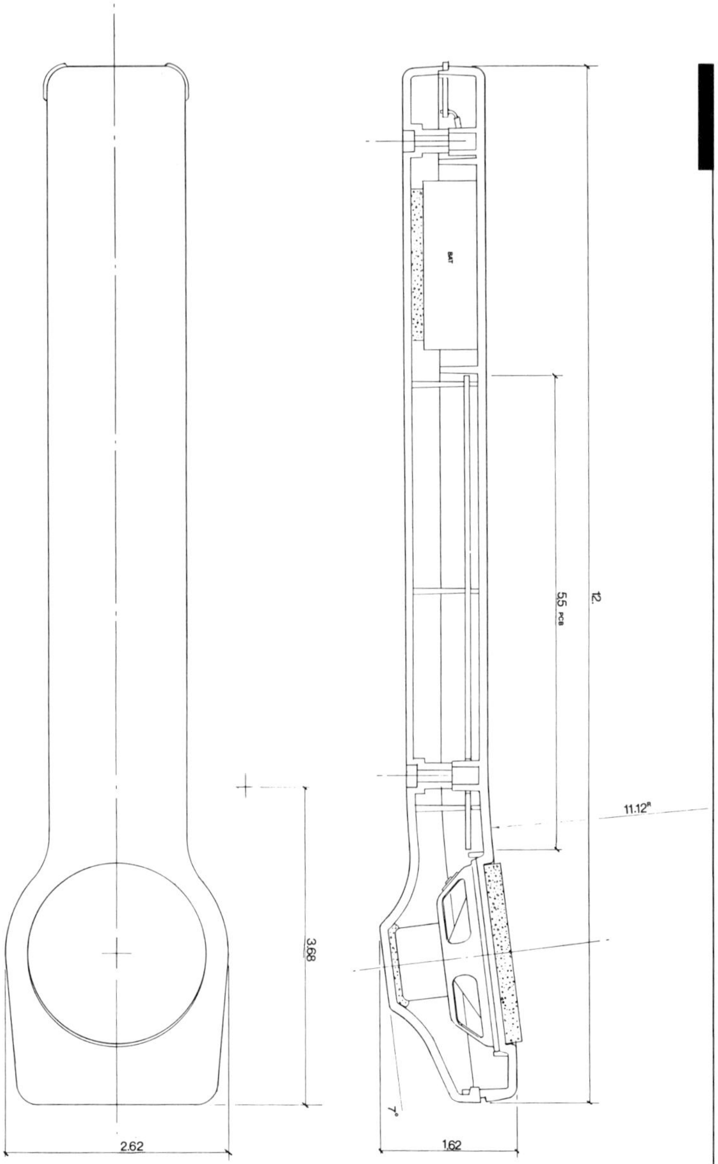

Form studies to determine shape and size.

The Design Problem: To design a handheld receiver for museums and historic sites that would be comfortable to use for long periods, easy to maintain, simple and attractive.

Form studies were made to determine the best shape and size. A design was selected for further tests that established handle shape and length. Esthetics and function were the foundation of the design development.

Ease of manufacture and servicing were as important to the client as convenience and appearance. Sound quality and storage for recharging were all considerations that were incorporated into the final solution.

Designer John Wistrand
Manufacturer By-Word, Inc.

Economy, lightness and sound quality were vital for this hand-held receiver used in museums.

IBM Executive Recorder 1979

Early concept drawings.

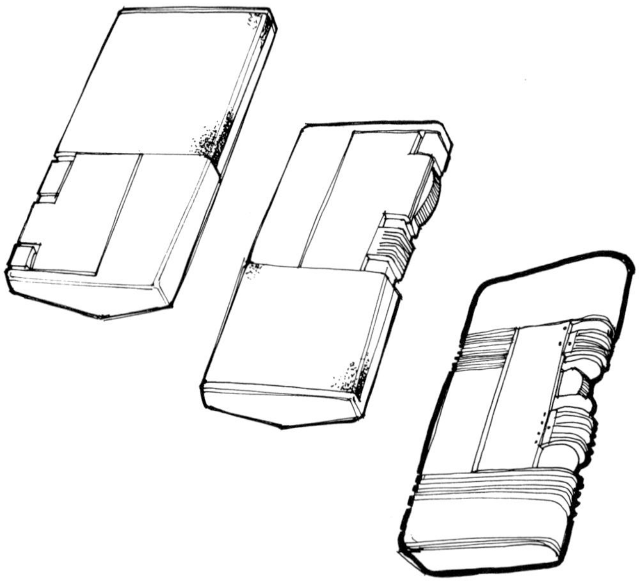

Three of the more than thirty models built to determine controls, speakers and microphone locations.

The Design Problem: To design a convenient, dependable and high quality pocket-size recorder with a simple, dignified appearance.

The small dictation unit was designed for one-hand operation; it is comfortable to use for extended periods.

Comparative acoustical performance tests against competitive equipment were conducted by IBM to assure high quality performance of the recorder's separate, built-in speaker.

Over 30 original models were built to determine the location of controls, speakers and microphone. The controls were designed to be operated by the thumb of one hand (index finger for left-handed users). A single, sliding main switch controls stop, review, reverse, forward and fast forward functions.

Design consultants Eliot Noyes & Associates
Industrial design Eliot Noyes
Project manager Gordon Bruce
Senior associate industrial design Allen Hawthorne
Graphic design consultant Paul Rand
Manufacturer International Business Machines Corp.
Industrial designer office products Vincent Giannotti, Jr.

Rubbermaid Stepstool 1975

Early concept sketch

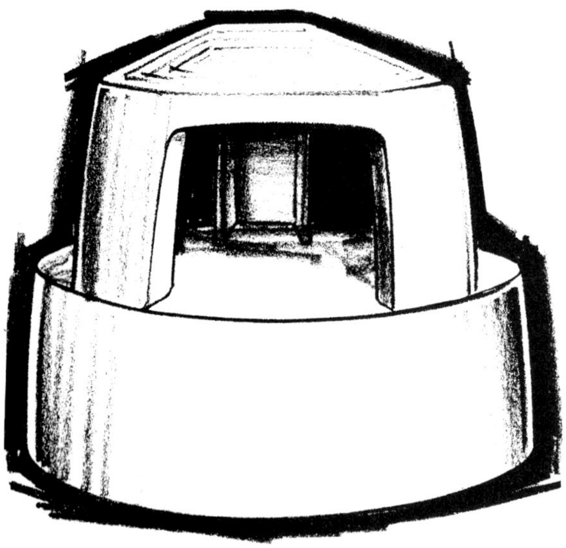

The Design Problem: To design a mobile unit for access to hard-to-reach areas that could also act as a temporary seat.

Research into competitive products indicated a need for a one-piece unit with a rolling base and fold-away back rest. The final design was simple, durable and lightweight.

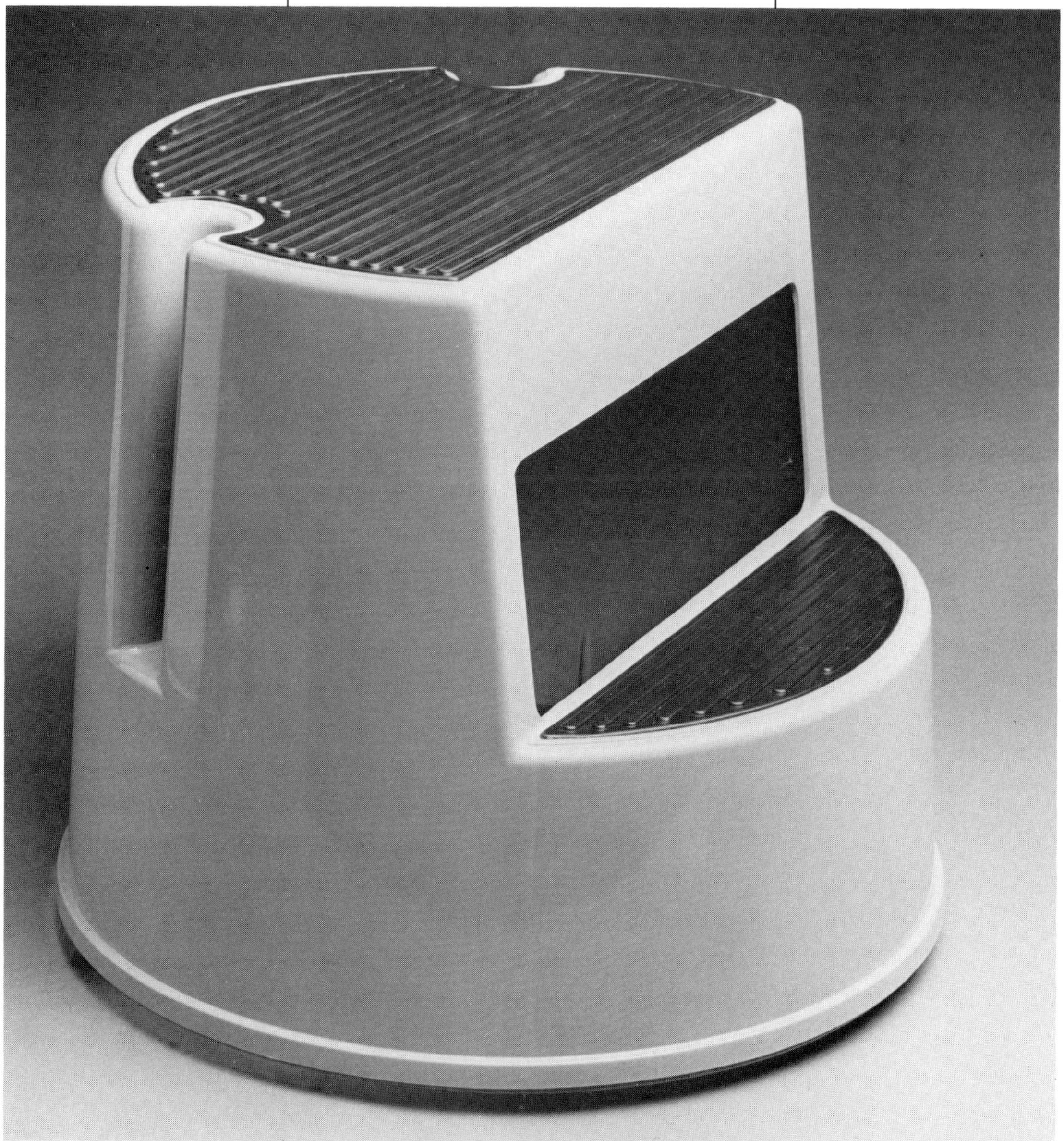

Manufacturer Rubbermaid Corp.
Senior project designer David A. Juergens
Staff designer William B. Raftery
Product engineering manager Jack Imhof

A mobile stepstool that doubles as a little seat is simple, durable and light.

Union Carbide Skipper Flashlight 1973

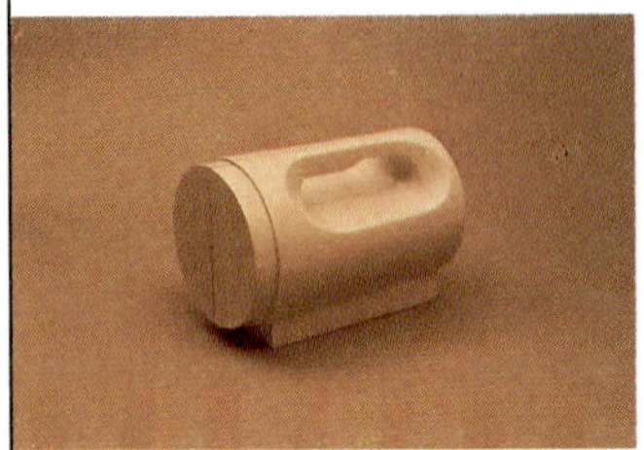

Solid form study for mold.

Clay mock-ups of lantern designs.

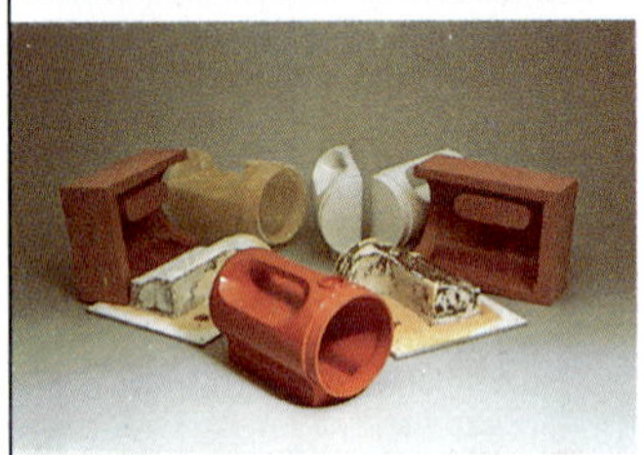

Components for first working model.

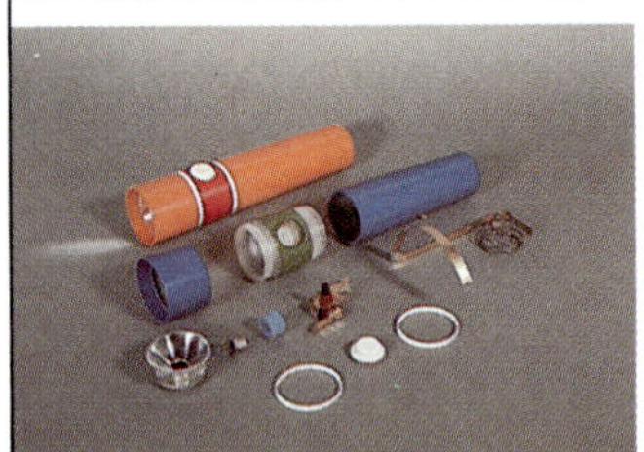

Color models for consumer testing.

The Design Problem: To develop a line of plastic flashlights that would achieve a notable functional advance: immersibility. In addition, the flashlight had to compete with existing metal flashlights in the marketplace.

The designer reviewed service problems and product use with consumers and distributors to determine possible acceptance of a plastic light. Metal flashlights had been considered "top of the line" products up to this time.

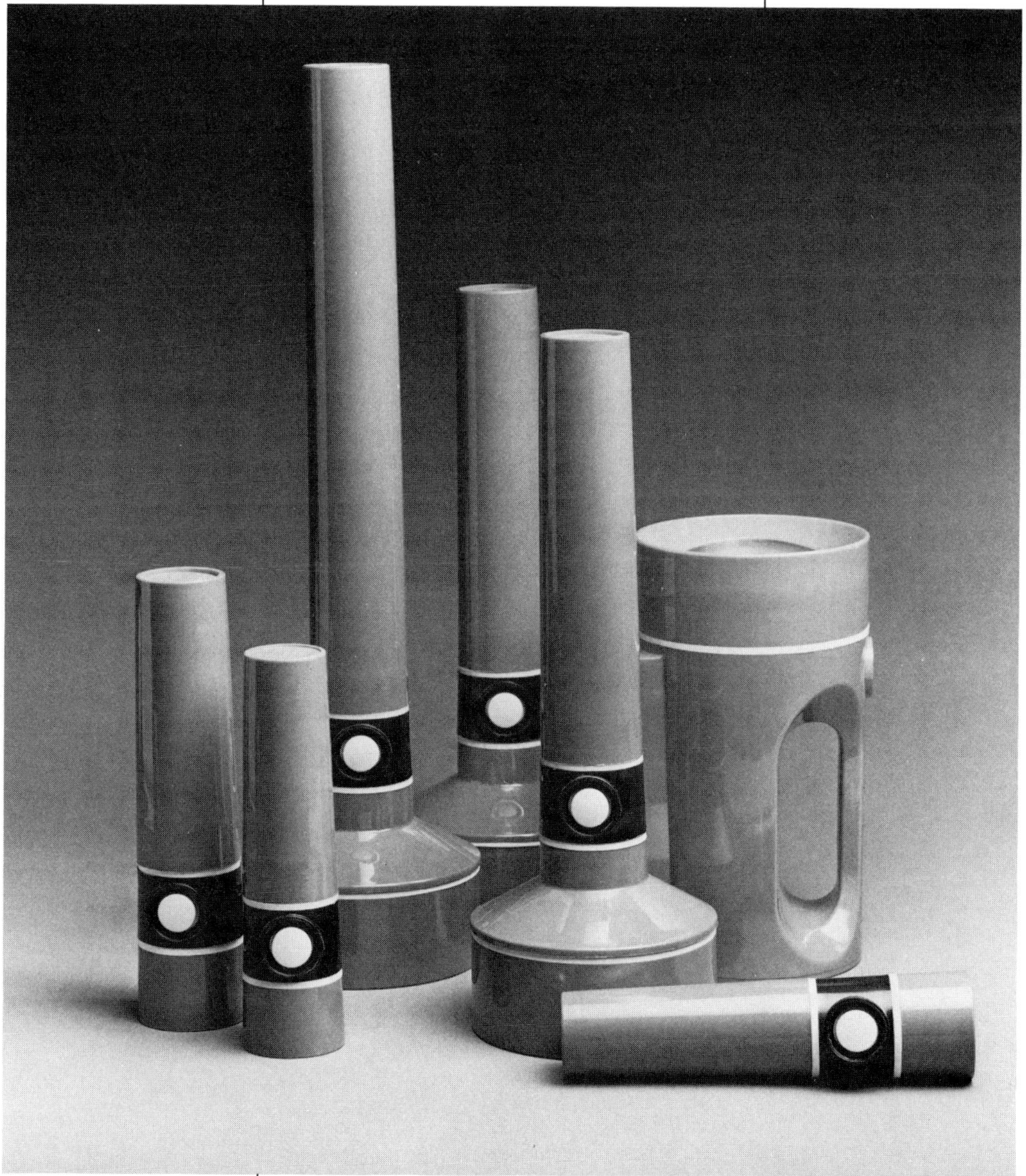

Design consultants David O. Chase Designs, Inc.
Designers David O. Chase, Martin V. Maloney
Manufacturer Union Carbide Corp., Battery Products Division
Designer Robert Brindley

The Skipper family (left to right): 2D cell, 2C cell, 3D cell; 3D and 5D cell searchlights, the lantern.

Union Carbide Skipper Flashlight 1973

After presentation of concepts, close cooperation with the client solved functional and technical problems. Extensive development work improved the switch mechanism and made it fully watertight.

The sporting market was a primary target for this new product: the innovative, floatable flashlight could withstand moist environments and could even be immersed in water.

Early working models.

Three colors used for introduction.

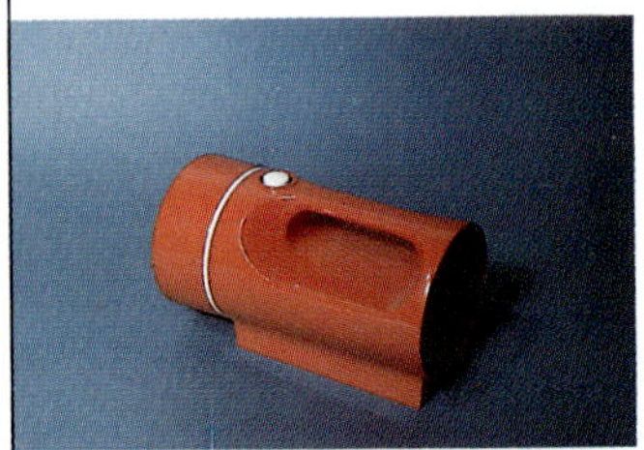

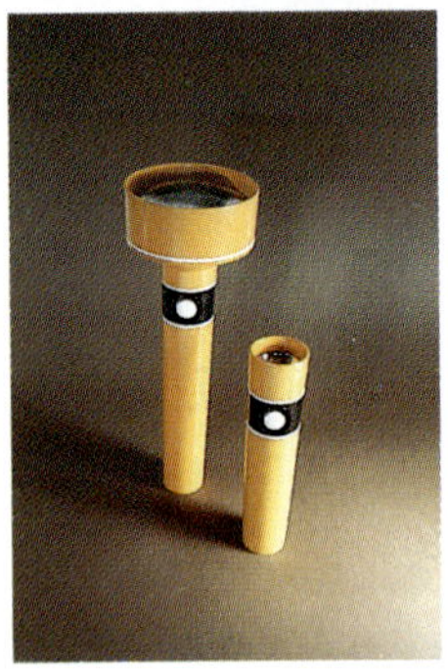

Texas Instruments Speak & Spell Teach & Read Programs
1977

The Design Problem: To develop an appealing educational product for the youth market that would aid in learning and spelling words.

Designed for grades one through eight, the level of difficulty is determined through a range of plug-in word modules. Market research conducted in focus groups concluded that both children and their parents considered such a product fun, that it provided value for money and kept children interested for extended periods of time.

Additional tests, including a series of design studies ranging from conceptual ideas to final appearance studies, influenced the direction for form, color and graphics. These were designed to help the product compete with toys in the market-place.

The selection of the product's "voice" was based on tests that determined the best voice quality, friendliness and user acceptance.

Designed for grades 1–8, the level of difficulty is determined through a range of plug-in word modules.

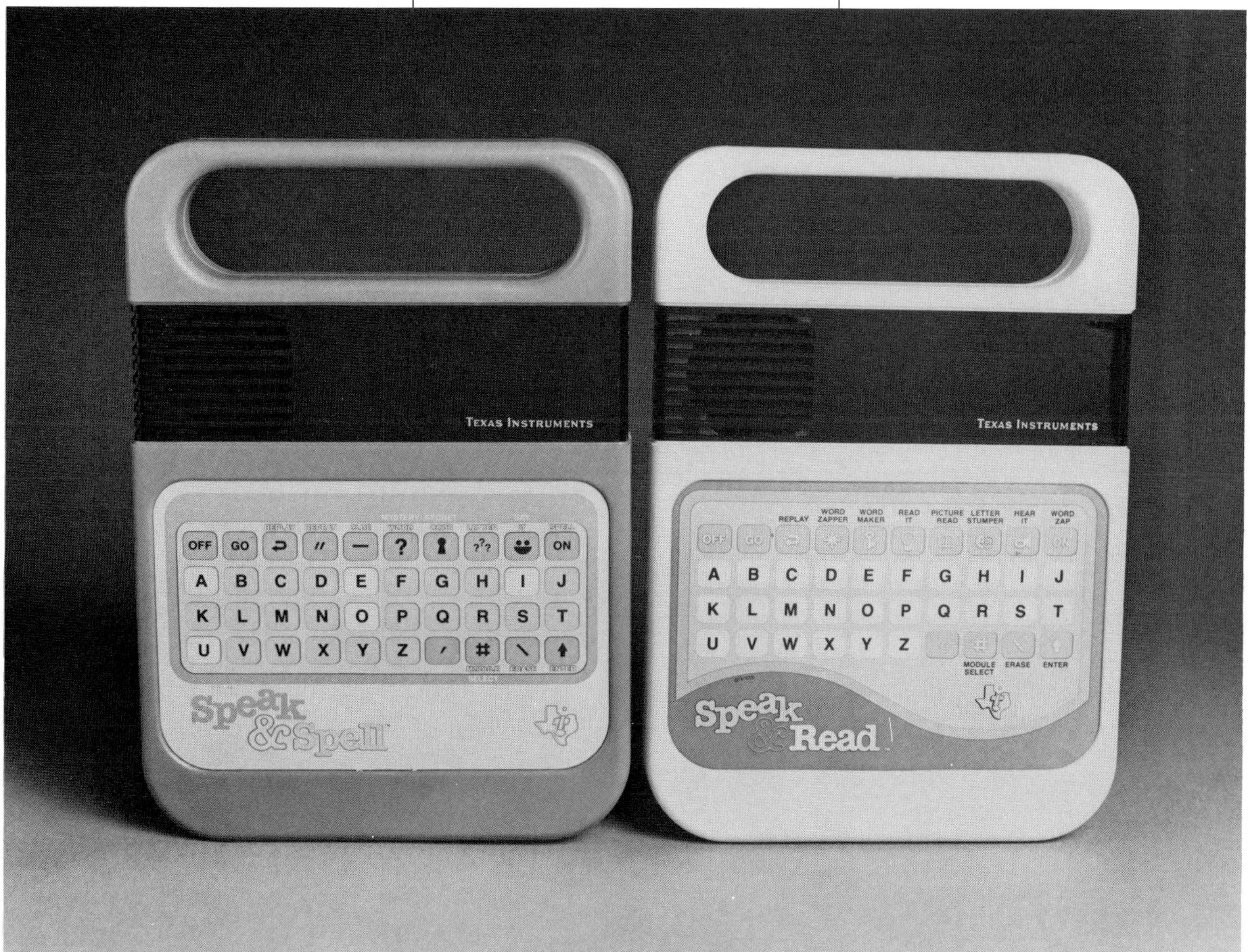

Manufacturer Texas Instruments, Inc.
Industrial design manager Eugene Sulek
Industrial designers W. Lawrence, R. Nesbitt
Engineering manager Paul Breedlove
Marketing manager Steven Weinstein

Amerac Precision Rowing Machine 1981

The Design Problem: To create a unique lightweight rowing machine capable of outperforming current rowing machines.

The 610 is fabricated of custom extruded aluminum, making it the lightest machine on the market, and portable (weighs 30 lbs.). The designer incorporated pivoting foot pedals for correct body to foot alignment during the complete rowing cycle and a comfortable, padded vinyl seat with a unique, ball bearing tracking system for smooth, easy riding.

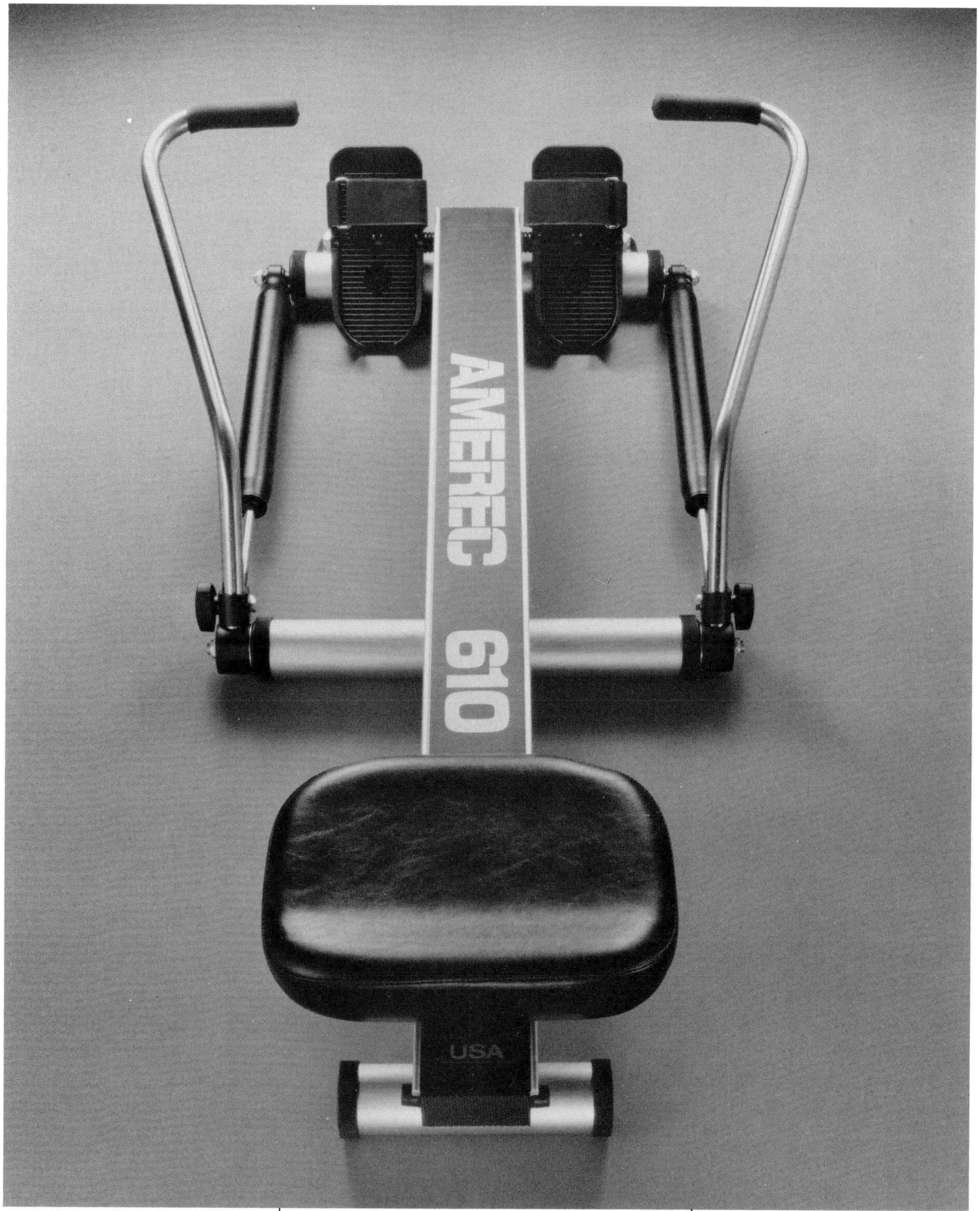

Design consultants Precor USA
President and designer David B. Smith
Manufacturer Amerac Corp.

The 610 is fabricated of custom extruded aluminum so that, at only thirty pounds, it is portable.

Cosco Go-Seat Child's Chair 1978

The Design Problem: To design an adjustable child's seat for use at the table, in a child's room or in a car.

The objective, as defined by Cosco's marketing department, was to design a safe and comfortable chair for children ages one to four. Early research and testing indicated the need to accommodate a child dressed in either lightweight or bulky clothing.

The design development was evolutionary. The process focused on two areas: the creation of a simple, sculptural form and the selection of the appropriate material.

An injection molded ABS plastic seat met the design criteria for a lightweight, non-breakable, easy-to-clean chair of the requisite strength that could be efficiently and economically manufactured. The adjustable metal base, developed by the designer, snaps into the bottom of the seat for ease of assembly. Self-leveling rubber feet increase stability and safety.

Raymond Spilman notes that, "the ultimate simplicity and functional form, developed with visual sensitivity into a practical application, is a clear expression of our approach to design tasks."

Design consultants Raymond Spilman Industrial Design
Designers Raymond Spilman, Robert Winters
Manufacturer Cosco Peterson, Division of Cosco
Engineer Clarence Hamilton II
Director R&D E. Keith Moore
Tool designer Don Gerkin
Marketing director juvenile products John Nolan

One objective was to accommodate a child dressed in light or bulky clothing. The molded, easy-to-clean seat has self-leveling rubber feet to increase stability.

Zelco Itty Bitty® Book Light 1982

The Design Problem: To design a lightweight book light that would work either on a/c current or battery pack and that would create a sense of whimsy while retaining a functional appearance.

Research into patent office files as well as discussions with major mail order companies and department stores indicated that the design should have a high quality look emphasizing lightness and compactness. Original design concepts had the look of a "tool" and would not trigger the buying impulse the designers were looking for.

Critical to the design was the development of a bulb that would produce a cool, shadowless, but brilliant white light.

Features that were incorporated into the final design included a flexible clip to accommodate varying book sizes and a rotating head that holds the bulb and angles for maximum light coverage. The final solution produced a pleasing design that invites the consumer to touch and use.

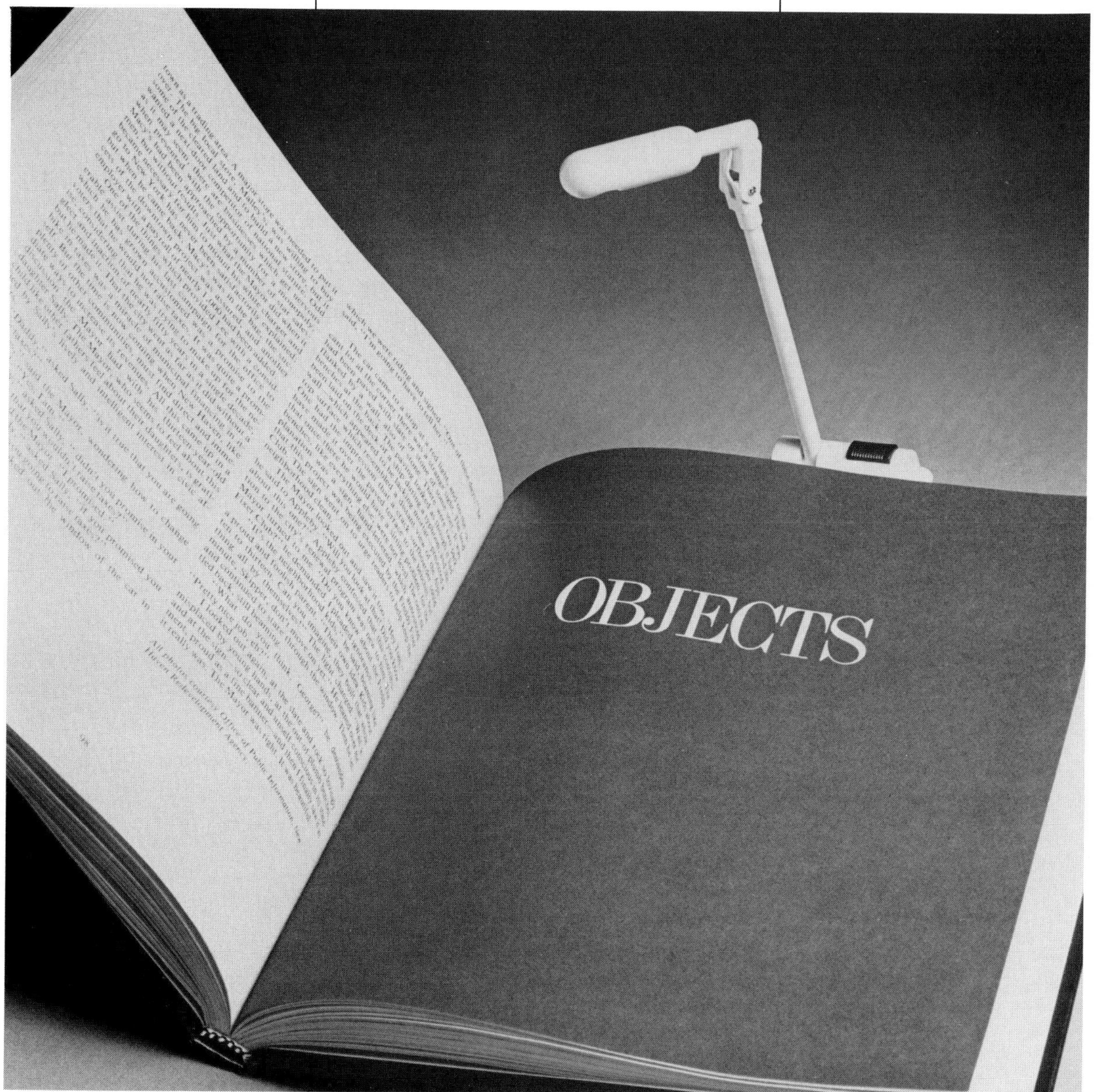

Manufacturer Zelco Industries, Inc.
Concept Noel E. Zeller
Designers Ben Zapolski, Fred Leung

Vital to the design was a flexible clip to accommodate various book sizes and a special bulb to provide maximum illumination.

★ **Grid Compass Computer** 1982

Early concept sketches.

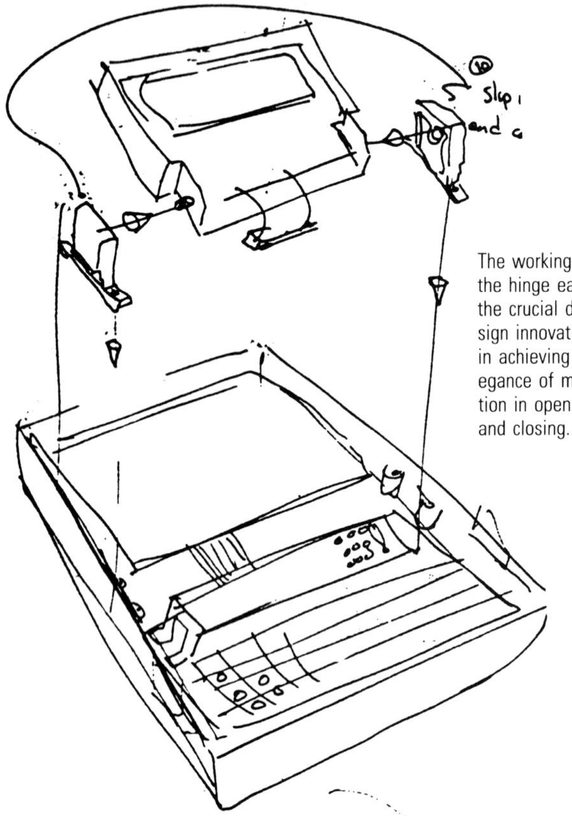

The workings of the hinge ear, the crucial design innovation in achieving elegance of motion in opening and closing.

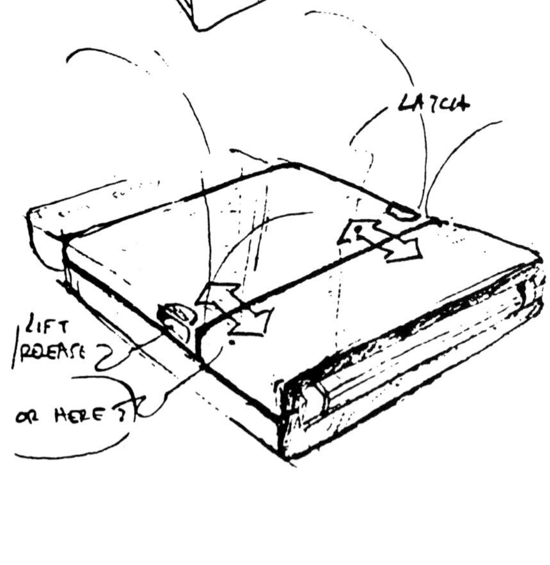

The Design Problem: To develop a powerful, truly portable, stand-alone computer for managers and professionals. The product concept, which was presented to the designers by the client, combined several new technologies that would make a truly portable computer possible.

The designers reported, "An appearance was sought to express the leading edge technology of the product, while remaining strikingly simple and elegant . . . to look international, belonging in a pan-world, rather than a national marketplace." One limitation was the lack of precedents.

The original concept was to design a product that when folded would fit within half the average briefcase, filling either the right or left side. Later the design team learned they were filling the wrong half of the briefcase! The next model reoriented the design (working with wood models) to fill the bottom half of a briefcase leaving room for papers to lie on top.

The use of magnesium for the computer's housing was innovative. Its strength, cooling qualities and lightness allowed for some complicated details such as the two hinge ears, the crucial element in achieving elegance of motion in opening and closing. Called the "backbone" of the entire assembly, the hinge ears support and allow the display to pivot and contain the channel for the data and power lines. They also support a torsion spring that prevents users from slamming the display closed. Appearance models of the new design tested the ruggedness of the unit.

Human factors considerations were important—knowing how the product was to be used, designing circuits and surfaces that would not burn people and finding a way to keep things out of the box. This was achieved by creating a deliberate scoop/pocket to collect loose briefcase objects, such as pens and paper clips.

The Compass is part of the Design Collection of the Museum of Modern Art.

Design consultants ID Two
Industrial design Bill Moggridge
Mechanical engineering Stephen R. Hobson
Manufacturer Grid Systems Corp.

The Grid owes its success to the fact that industrial design began before any other activity was undertaken.

Steinberger Electric Bass Guitar 1981

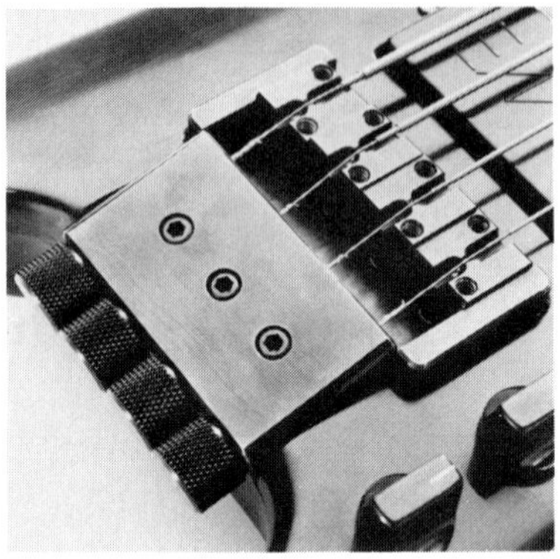

Steinberger developed a simplified and accurate tuning mechanism and a method for quickly changing and re-using strings.

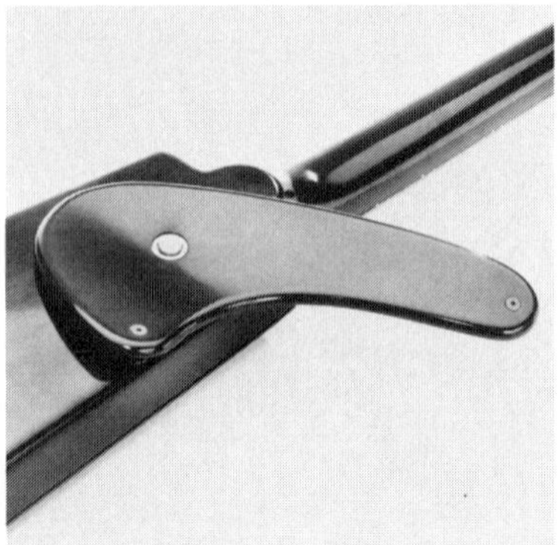

The strap support plate pivots at a single point on the back of the bass at the center of gravity, allowing the instrument to be played at any angle in perfect balance with access to all 24 frets.

The Design Problem: To improve the sound and playability of the electric bass guitar, using current technology. A major objective was to develop an esthetic that would reflect the new performance level without changing the playing method.

The electric bass guitar was first introduced in 1951 by Leo Fender, who virtually monopolized the bass market for thirty years. The Steinberger is the first major redesign of the electric bass; the instrument was redesigned to optimize the relationship between the player and his instrument, not for the sake of appearance or novelty.

Early in the project, the designer realized that the inherent neck-heavy imbalance could be rectified by moving the tuning mechanism from the neck to the body of the instrument. He reduced the over-all dimensions to increase comfort, both for use and portability.

Sound in a musical instrument is the most important design factor. The first models in wood had a dull, unevenly sustained sound. Testing proved that sustain and brilliance improved when the designer covered the entire instrument with several layers of fiberglass and resin to increase its weight and stiffness. As experiments continued, Steinberger discovered that a high performance instrument must be as stiff as possible within a practical weight limit.

The final product, which Steinberger manufactures and markets through his own company, is made entirely of a one piece carbon graphite fiber and glass reinforced epoxy. The strong, rigid construction provides tone clarity and long sustain and opens a new range of sound to the user. The neck will not bend or warp even under highest string tension, and the instrument is unaffected by temperature changes and humidity.

The unfamiliar but compelling look of the Steinberger Bass is a direct result of its mechanical design and the materials used.

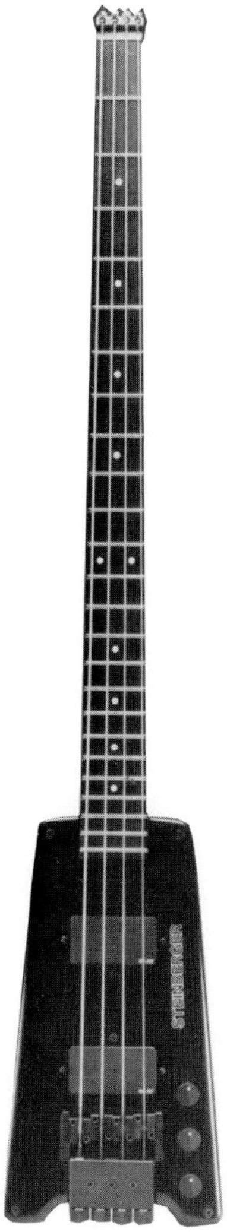

The unfamiliar shape is a byproduct of mechanical innovation, extraordinary balance, and the one piece graphite-glass reinforced epoxy material.

Manufacturer Steinberger Sound Corp.
Designer Ned Steinberger

Moss Starlet Backpacking Tent
1982

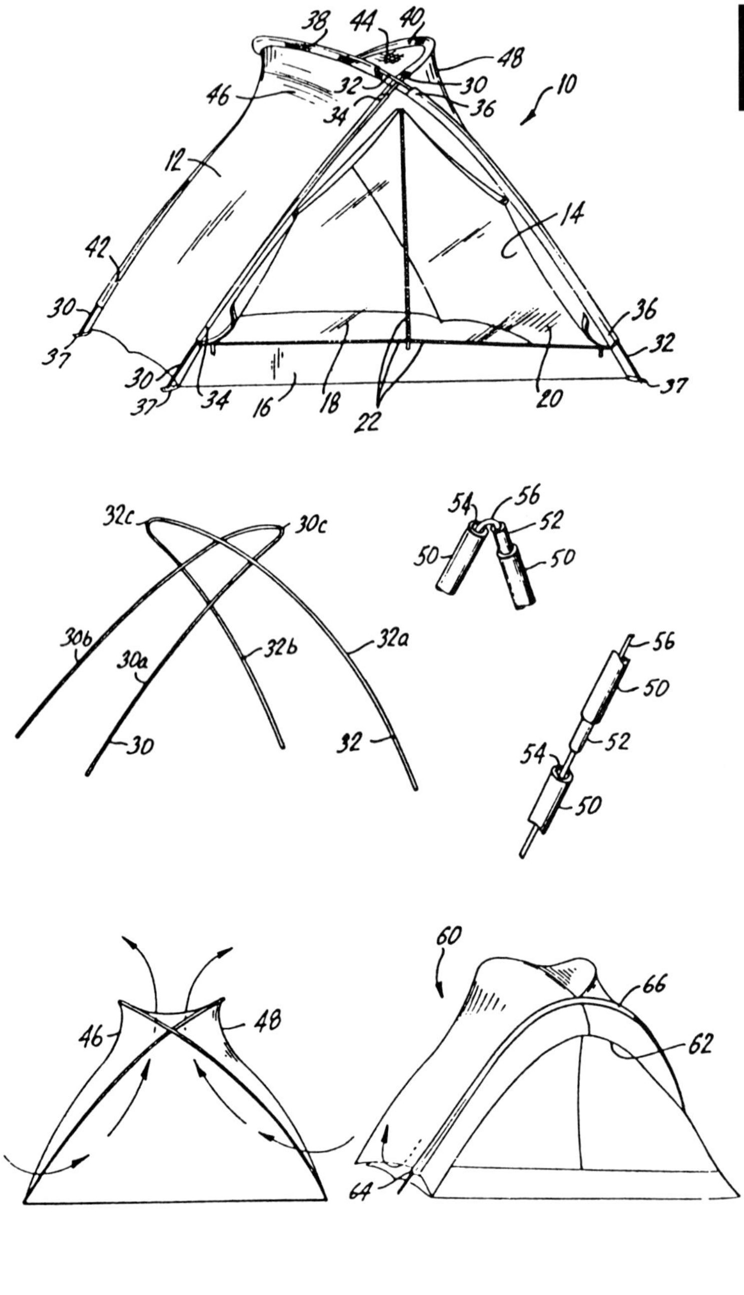

Patented structural forms use a tensioned fabric membrane that interacts with a curved framework.

The Design Problem: To design a low-cost lightweight backpacking tent for two persons that would compete favorably in the market with imported tents. The tent had to be easy to erect and stable under wind loads.

Moss Tent Works is a professional design group with production and manufacturing capabilities. All Moss patented structural forms utilize a tensioned fabric membrane, interacting with a curved framework, to realize the optimal strength of the entire structure.

In the Starlet, intersecting curved poles with fabric under tension provide a shape that deflects the wind and a stable base that eliminates extensive staking and permits the tent to be free-standing. The flared top, closed by netting, increases ventilation and serves as a skylight.

The designer says, "At some point in the design process, the very restrictions of the problem pressed out a creative solution which resulted in a streamlined, aerodynamic form which is esthetic, extremely stable, and spatially functional."

Design solution yielded a streamlined, aerodynamic form that was beautiful and functional.

Manufacturer Moss Tent Works, Inc.
Vice president senior designer Bill Moss
Designers Jan Janville, Susan Willer, Patricia Lees
Tent development specialist Ross Warren

Kodak Disc Camera 1982

The Design Problem: To design a camera around an innovative film disc. Conceptually, it was founded on a concern for the user and the desire to create a new system of decision-free photography.

Research into why and how people take pictures played a major role in early information gathering. Examination of thousands of customers' prints revealed the need for decision-free features in the new camera: automatic flash activation and recharging, trigger location to minimize camera shake and flash location to reduce the chance of finger-over-flash and red eye.

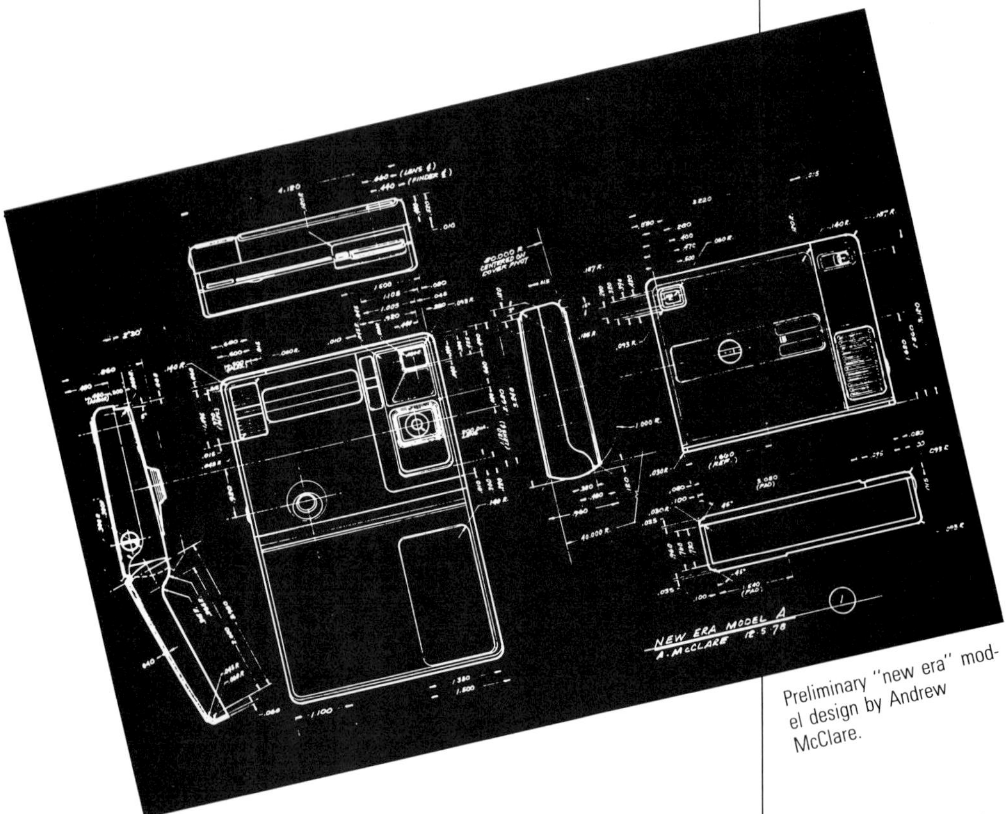

Preliminary "new era" model design by Andrew McClare.

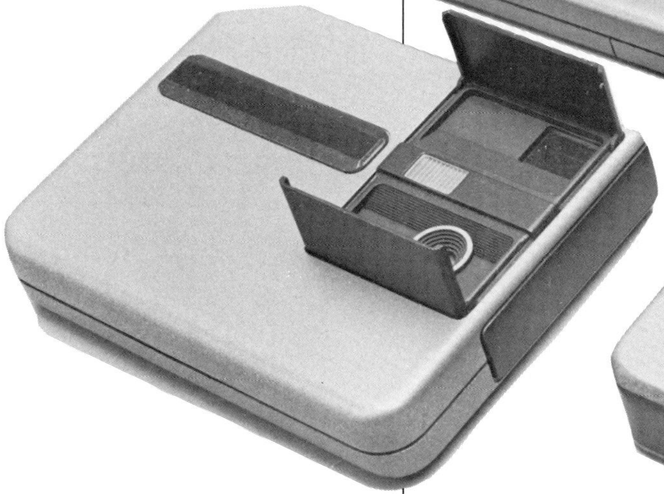

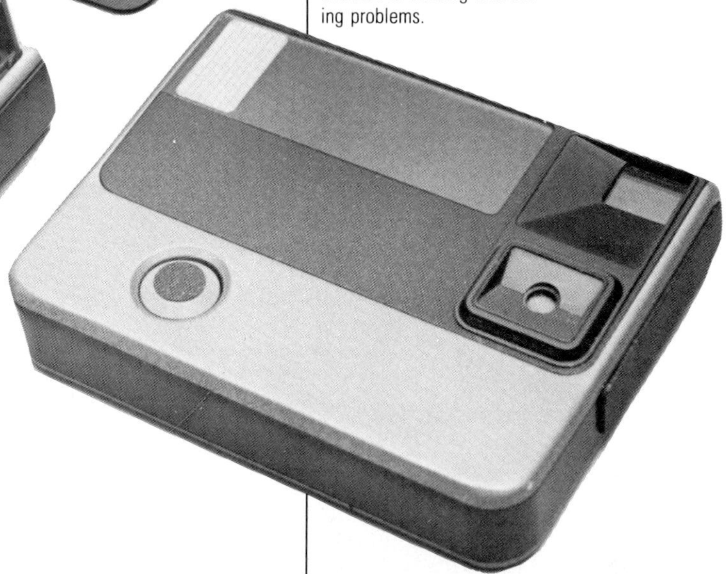

Early development models based on a horizontal format with the film loaded at the bottom and the camera held like binoculars was rejected because of holding and aiming problems.

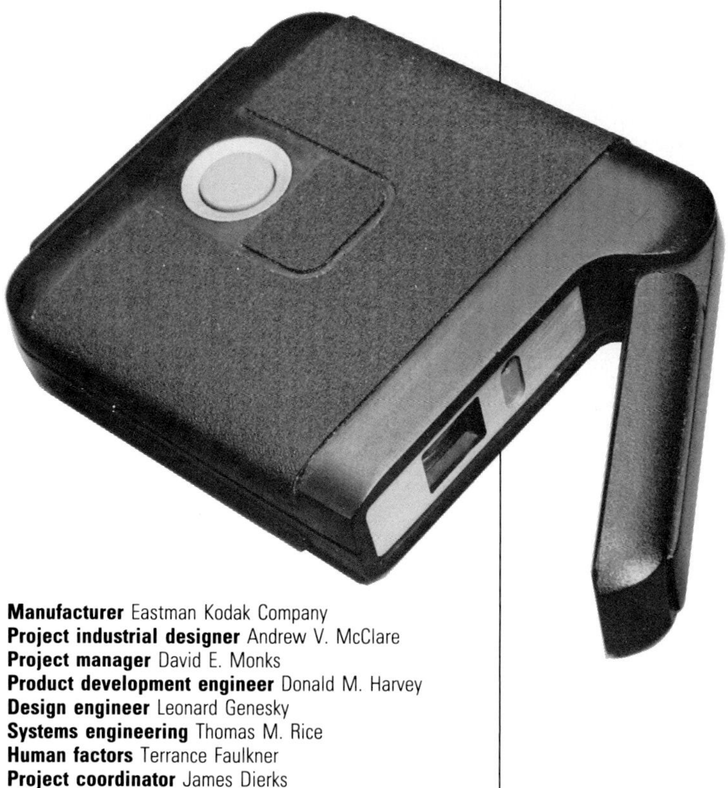

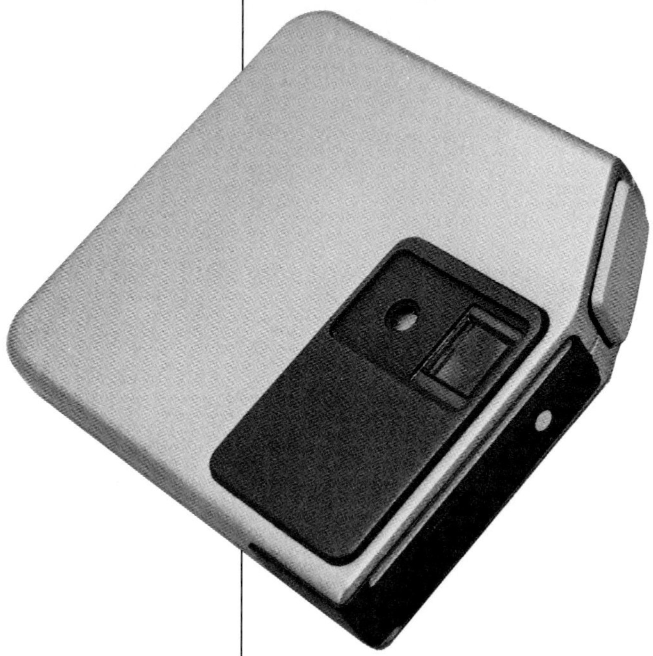

Manufacturer Eastman Kodak Company
Project industrial designer Andrew V. McClare
Project manager David E. Monks
Product development engineer Donald M. Harvey
Design engineer Leonard Genesky
Systems engineering Thomas M. Rice
Human factors Terrance Faulkner
Project coordinator James Dierks

Kodak Disc Camera 1982

Discussing his experience in designing the Disc camera, Andrew McClare states that, "In broad terms my involvement responded to a lesson taught by the history of art: 'Design follows invention but precedes innovation.' Specifically, my efforts created not only the surface appearance but, through working with others, the form and embodiment of a new photographic era."

The marketing department's approval of the product appearance was pivotal to the development process. These renderings were made in response to marketing's rquest for "elegance" and illustrate the use of line and color to communicate options.

Detailed models were developed to examine form, component layout, lens covering techniques and the impact of these factors on holding preference.

The final product line for the disc camera that introduced "decision-free" photography.

Savin Plain Paper Copier 1980

The Design Problem: To design the world's smallest plain paper copier, which would appeal to both the traditional business buyer and the home consumer.

The first step in the design was to determine the shape. It was decided to emphasize the small size, "cuteness" and almost toy-like quality of the copier. The designers state, "It had to be simple and precise looking with fine detail incorporated in order to play on the visual characteristics of expensive, high technology consumer products like cameras and stereo components."

The second step was to make the "how to use" functions clearly understandable and appropriate to a simple machine. Each control group is separated by color. There are five short lines of instructions and a very large, brightly colored knob on the paper tray. These are the only details necessary for operating the unit.

Overall coloring had to be neutral to fit into a variety of interiors, yet there had to be some bright coloration for contrast. The designers chose orange. The form, base don simple geometric shapes, is straightforward, yet the details—colors, textures and graphics—appeal on an emotional level, making the copier a satisfying tool to use on a daily basis.

The small "footprint" of the table-top model makes it ideal for cramped office situations.

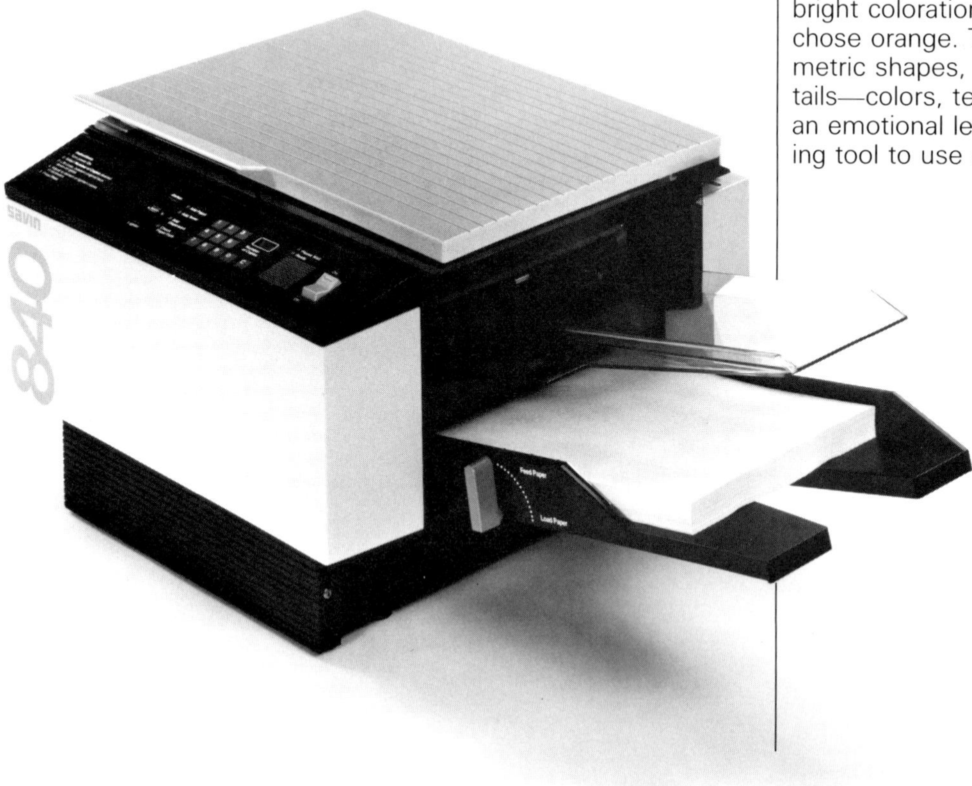

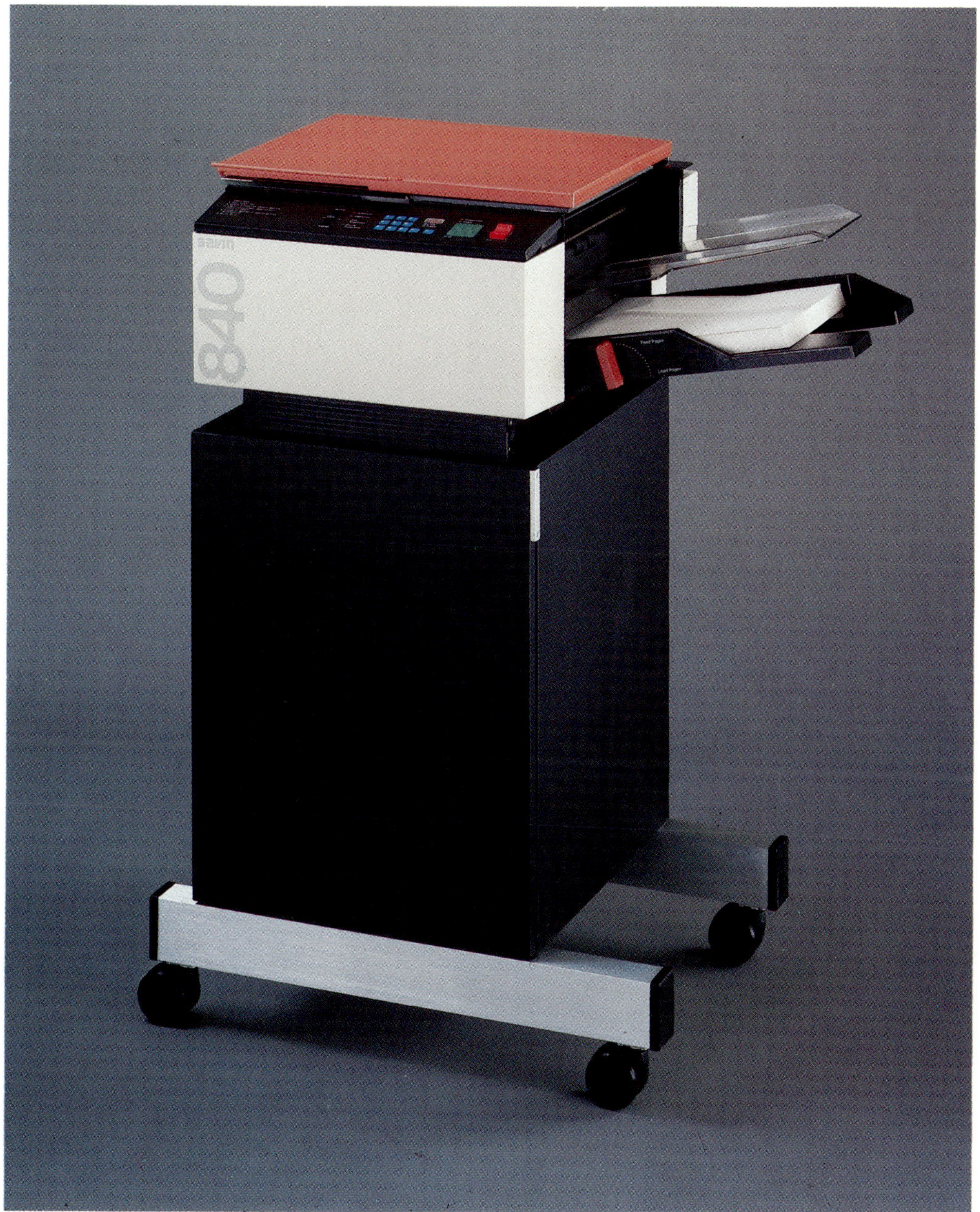

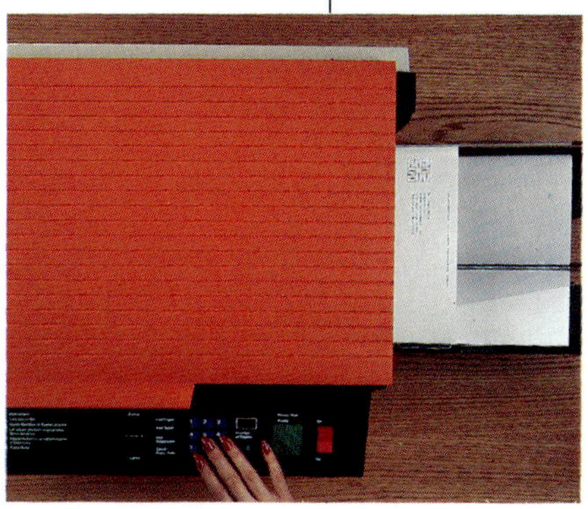

Each control group is separated by color; there are only five short instruction statements, and the knob on the paper tray is large and brightly colored.

Design consultants Plumb Design Group
Designers William Lansing Plumb, Andrew Serbinski
Design support Melinda Courrier, Steven Bellofatto
Manufacturer Ricoh Co. Ltd.
Client Savin Corp.

Corning Digital pH Meter 1983

Idea drawings of the instrument's assembly.

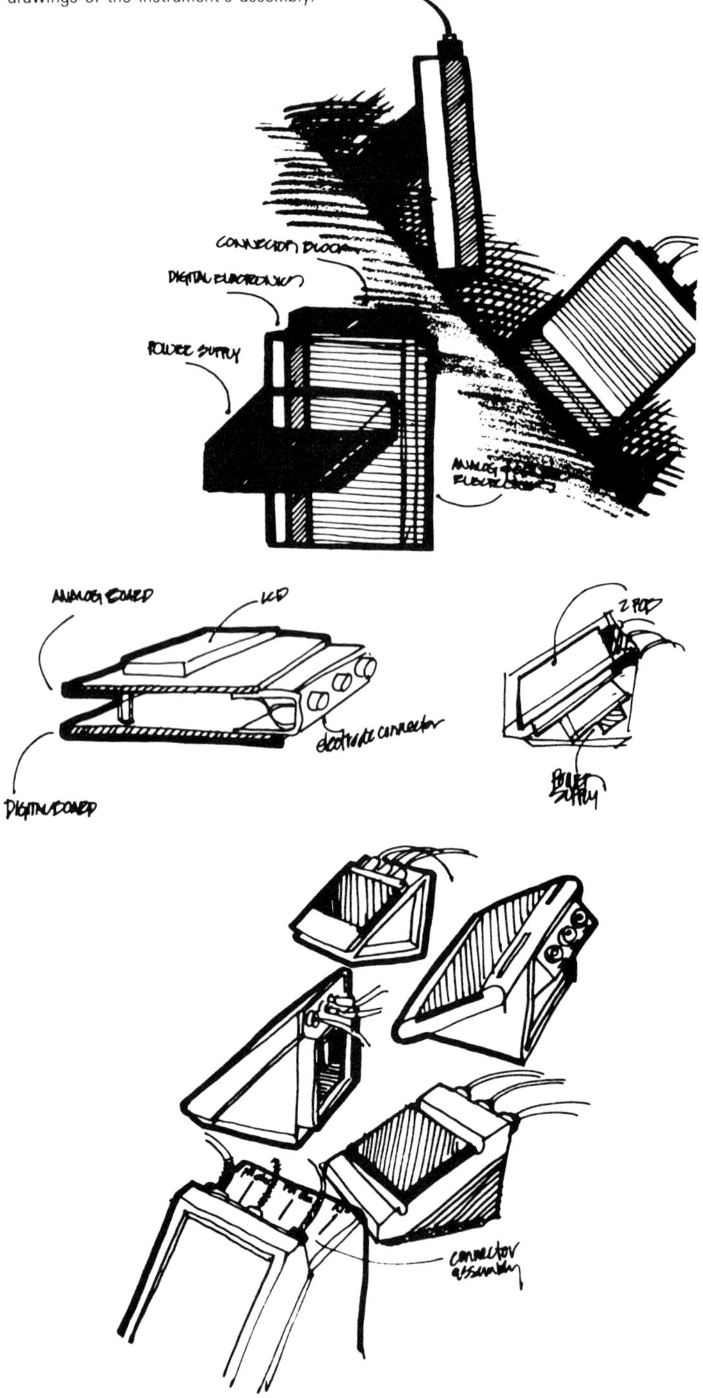

The Design Problem: To fill the need for a technically competitive, cost effective replacement for an existing line of scientific/industrial pH meters.

The pH meter measures acidity/alkalinity, ionic strength and the temperature of both aqueous and solid samples. It is used in medical and scientific research, food processing and industrial process control. Corning's pH meter line had changed little in recent years. Several units were over ten years old, based on older components and manufacturing methods and were no longer profitable to produce.

Ease of use and simplicity were of primary concern in the operational design and external appearance of the new product line. An "Operational Sequence Diagram," a blueprint of instrument usage and user interaction, permitted evaluation of the operational design. This research resulted in a simplified format for the pH meter that essentially guides the user through a measurement sequence by engaging him in a series of questions and statements.

As the product became a tangible set of components, the designer's conception of the instrument "as a book" was realized. The control panel and printed circuit boards are mounted parallel to one another "as pages" and propped up "for reading" by a structure containing the power supply. Thus, the exterior form implies the interior content.

A prime element of Corning Medical's development methodology was involvement of the designer in each step of the process: marketing specifications,operational sequences, component selection, structural design and human engineering. In this way the designer acts as a communicator between business sectors, translating marketing, manufacturing and customer needs into a common focus.

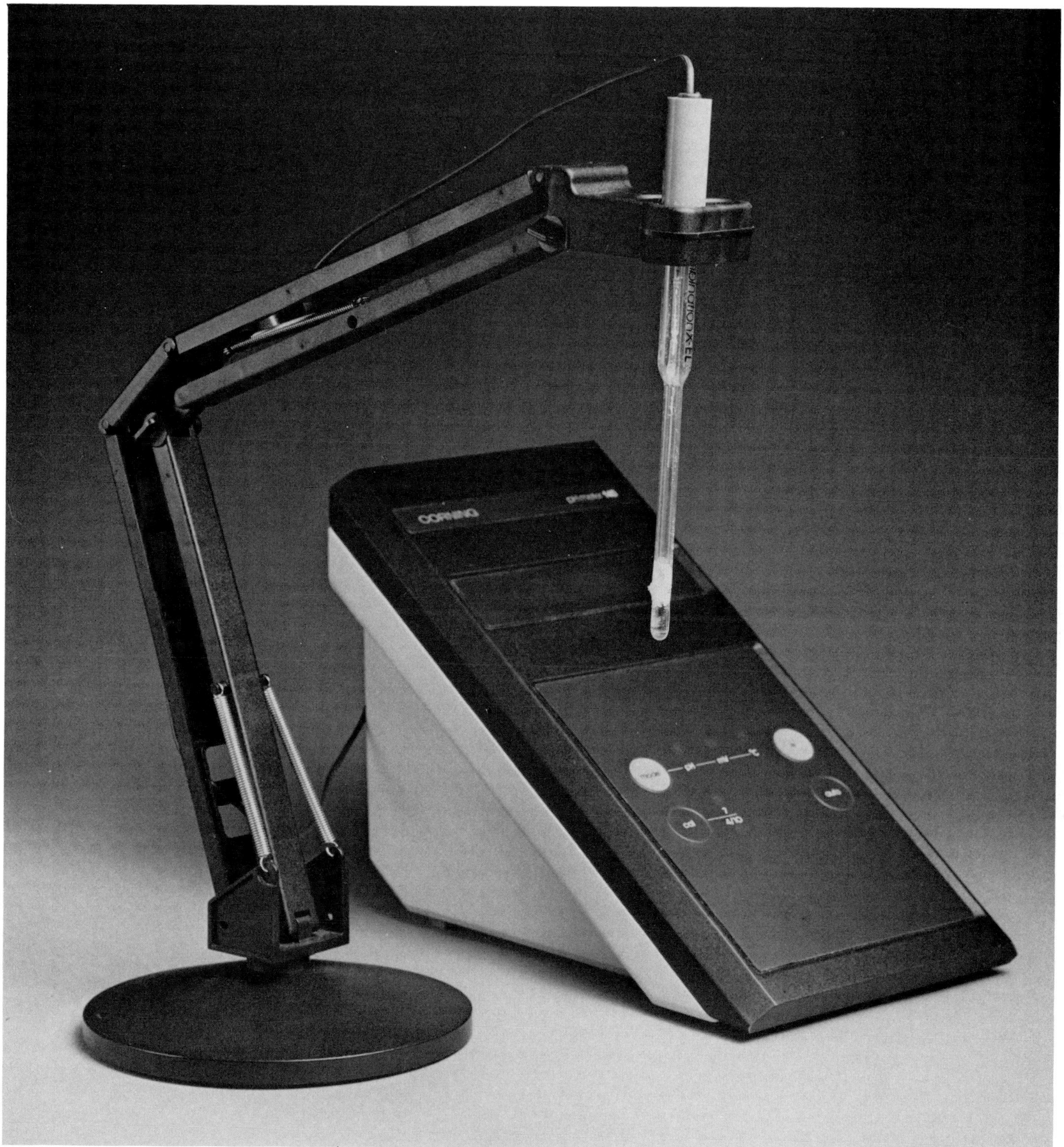

Manufacturer Corning Medical and Scientific, Div. Corning Glass Works
Product designer Scott Mathis
Project manager Chris Orlopp
Mechanical engineering Derek Archer

Control panel and printed circuit boards are mounted in parallel and propped up for reading like a book.

NCR Tower Microprocessor
1983

The Design Problem: To develop an office system composed of modular units with a distinct product personality.

The Tower is a medium range data processor suitable for small business use. Designed in-house by NCR, the Tower reflects the over-all Corporate Appearance Program developed by Richardson/Smith, consultant designers in conjunction with NCR's design management team. Dean Richardson explains that today the decentralization of the modern corporation requires an over-all guiding strategy.

Richardson/Smith works in advance of specific product design, acting as concept strategists. The Corporate Appearance Program, designed to create a personality for NCR products to the end of the decade, is motivated by human factors considerations that are the source for "the visual and tactile substance of the product."

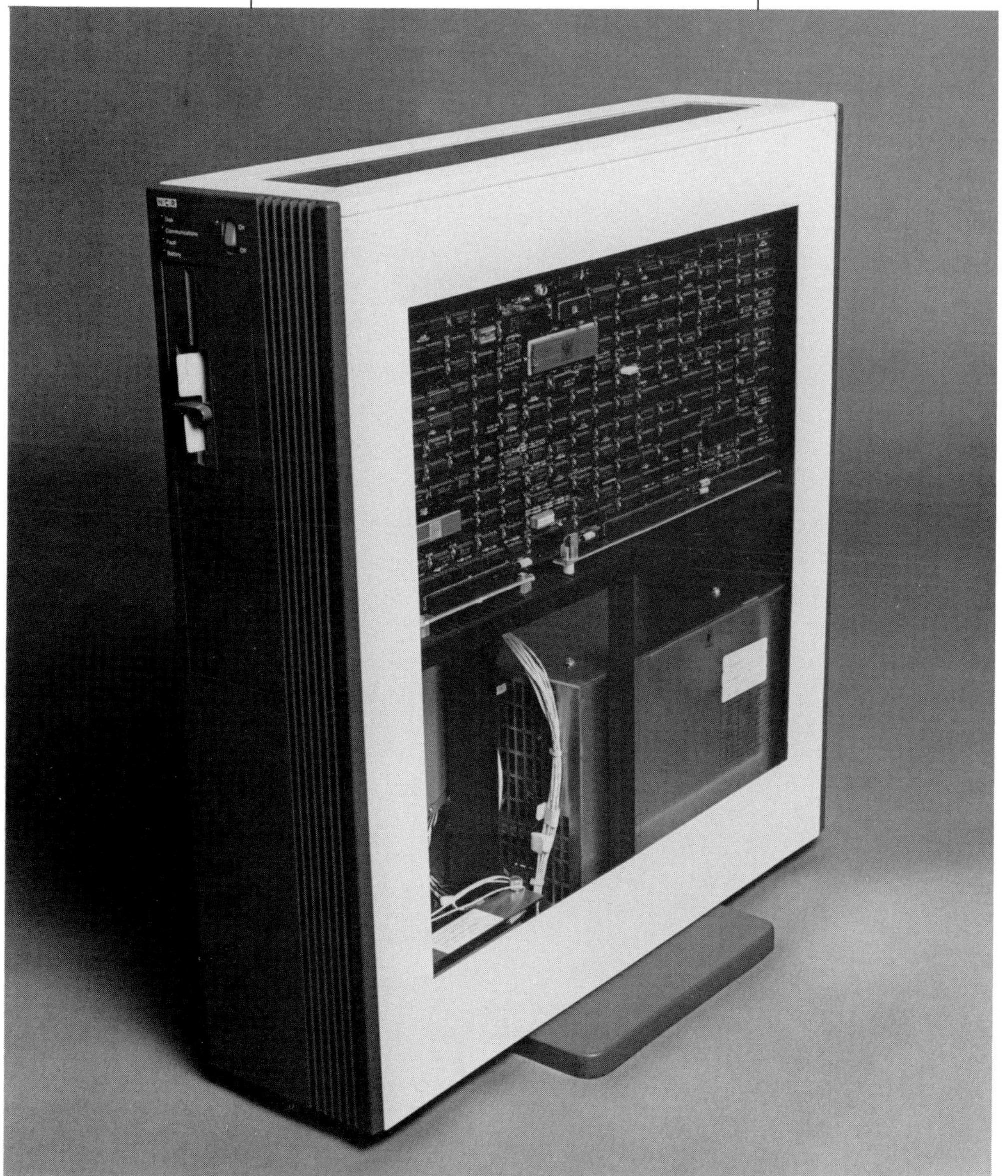

Design consultants Richardson/Smith, Inc.
Manufacturer NCR Corp.
Designers James Barnes, Edward J. Roswog

This medium range data processor, about four feet high, is designed to conform to NCR's Corporate Appearance Program developed by Richardson/Smith.

Convergent Technologies WorkSlate 1983

Concept sketches reach for a new form to organize an array of tasks.

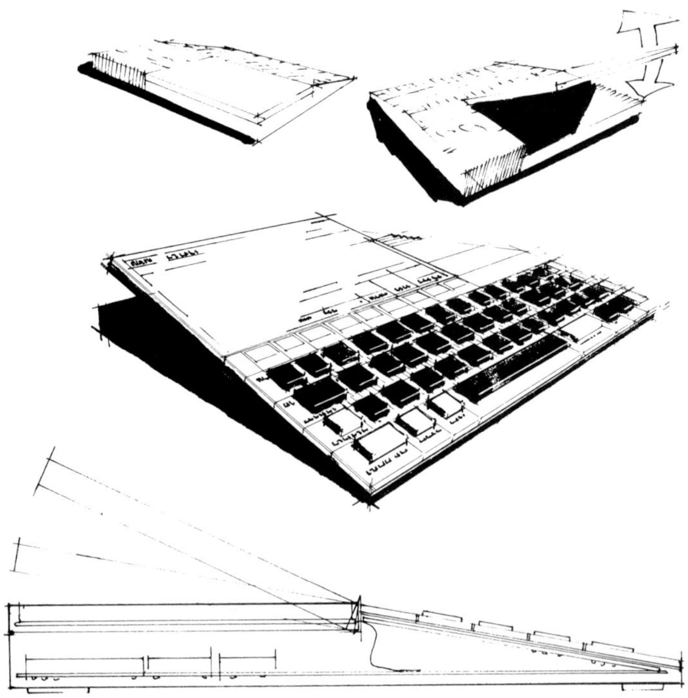

The Design Problem: To develop an electronic tool better suited than a personal computer to the requirements of business people.

The designers surveyed business people, both users and non-users of computers, to determine how and why they did or did not use computers. They analyzed the tasks business people did on a regular basis and examined their tools. Results indicated that people found computers forbidding and difficult to understand and that the tools most frequently used were paper, pencil, calculator, dictating machine, calendar, telephone and telephone book. These tools were not confined to the office desk.

The WorkSlate was then designed, not to have the appearance of a computer, but to reflect a new kind of product, based on the concept of an electronic work-sheet. It is note pad size, 8½ × 11, is easy to carry, weighs less than 3½ lbs. and is fun to use. It does not require computer language.

Innovative features include Taskware™ tapes, which store prebuilt "worksheets" (personal tax, portfolio analysis, cash management) and a 30 minute "Teach Me" tape. The user can prepare financial forecasts and cost estimates and also use the WorkSlate as a tape recorder, calculator, speaker phone and connection to other computers. The battery powered unit has a typewriter keyboard, a built-in digital and audio micro-cassette recorder that stores dictation or worksheets and a small, battery powered, multi-colored type printer.

The product reflects a new kind of configuration based on an electronic worksheet.

Design consultants Matrix Product Design
Industrial design Mike Nuttall
Manufacturer Convergent Technologies
Engineering development Charlie Florella, Bill Burnett

AMERICAN

3

FORM

DESIGN

CLASSICS

The Museum Watch 1961

The Howard Miller Museum Clock 1961 10/11

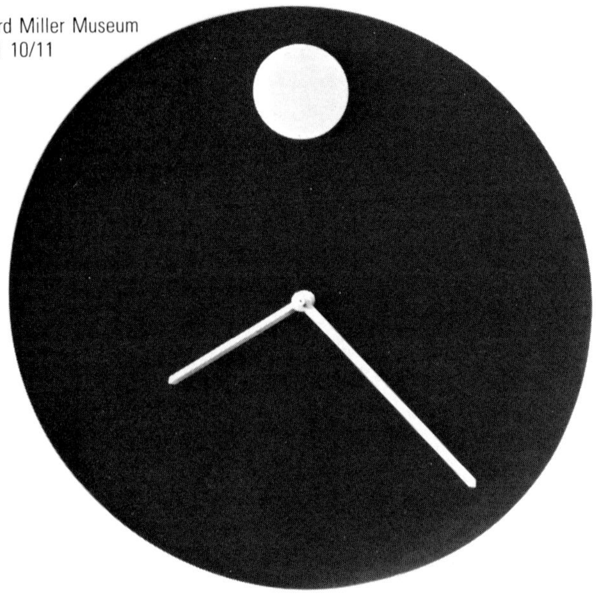

Nathan George Horwitt carried the simplification of the clock to utmost simplicity with only three indices.

The Design Problem: To relieve the clutter of the watch dial and to facilitate faster recognition of time.

Nathan George Horwitt describes himself as a "functionalist designer with little understanding of decoration. To offset the clutter on the face of a watch, I carried the simplication of the numberless dial to three indices—a lunar dot at the top to mark the twelve, and the two hands."

The watch first appeared as part of the Museum of Modern Art's Design Collection and later was licensed to Movado.

Edward Steichen wrote to Horwitt of his design: "I believe your design for the face of a watch is the only really original and beautiful design that I have ever seen. It places the element of design on an esthetic basis parallel to that of a painter's work."

The Howard Miller Clock is also part of the Design Collection at the Museum of Modern Art.

The Howard Miller Museum Clock 1961

The Howard Miller Clock is also part of the Design Collection at the Museum of Modern Art.

Designer Nathan George Horwitt
Manufacturer Howard Miller Clock Company

The watch first appeared as part of the Museum of Modern Art's design collection and was later licensed to Movado.

Designer Nathan George Horwitt
Licensed to Movado Watch Company

Lightolier Lytegem Hi-Intensity Lamp 1965

Advanced concept drawings.

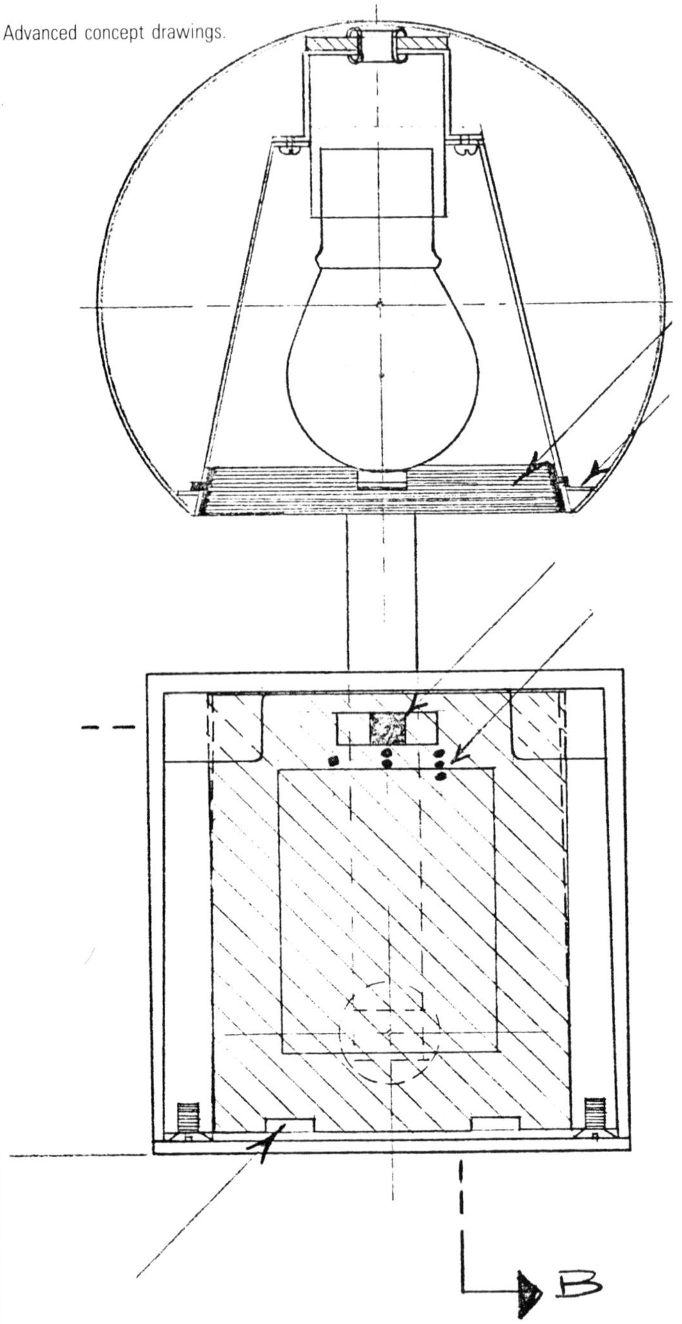

The Design Problem: To design a small, hi-intensity light to compete with similar products on the market.

Until the Lytegem appeared hi-intensity lights were used only in workshops and garages. This new light, to be used in homes, had to be stylish and attractive.

Michael Lax worked closely with the engineering staff at Lightolier who developed the telescoping arm and light shield. The base houses the transformer: the shade is a ball attached to the base by a telescopic swivel arm that extends 14 inches. The lamp is part of the Design Collection at the Museum of Modern Art.

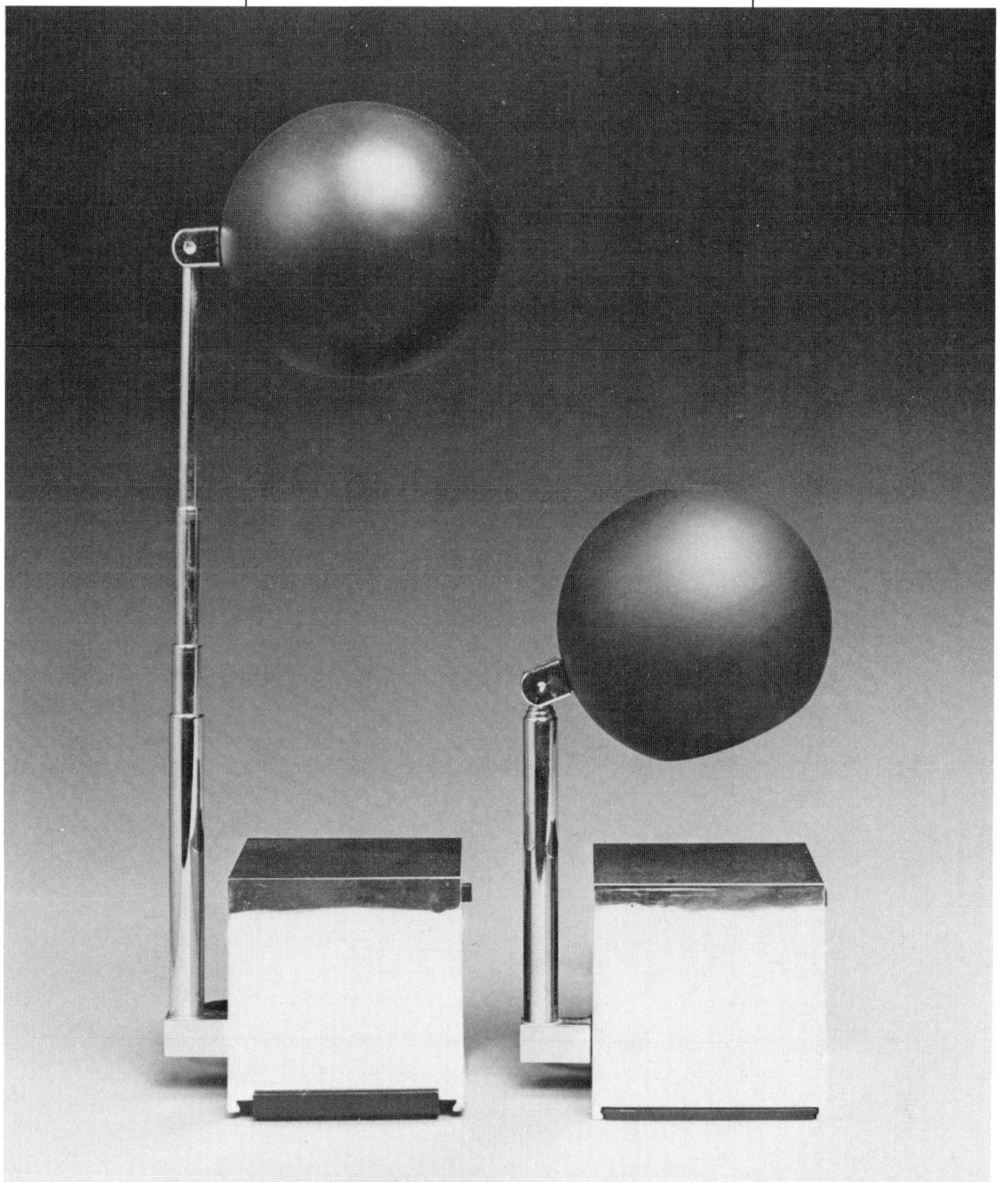

Designer Michael Lax
Manufacturer Lightolier, Inc.

Designer Michael Lax worked closely with Lightolier engineers to develop the telescoping arm and light shield.

Hartman Reflector Floor Lamp 1967

The Design Problem: To design a lamp that provides a versatility of light patterns.

The rotating, shielded light source helps to distribute light and to define or organize a space. The minimal base makes it convenient for placement under furniture.

Built of solid brass, all parts are cut, ground and polished by painstaking hand operations. This lamp, seen here in a nickel finish, is part of the Design Collection at the Museum of Modern Art.

The rotating shield helps to distribute light to a defined area.

Manufacturer Cedric Hartman, Inc.
Designer Cedric Hartman

Block China Transition Porcelain Dinnerware 1968

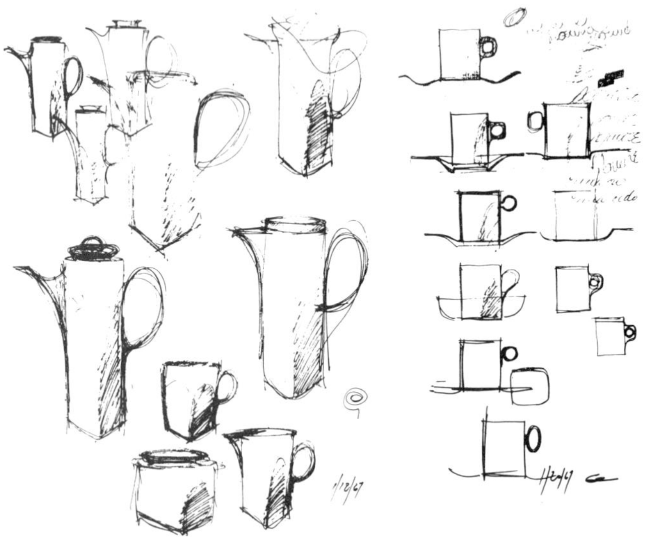

Concept sketches by Gerald Gulotta.

The Design Problem: To design casual porcelain dinnerware of substantial weight and stacking and nesting features that expressed a style beyond the purely utilitarian.

Casual market research had indicated that the consumer was eager for the unexpected. The design motivation was to create something timely rather than timeless.

The dominant design objective was to develop interrelated forms that would stack vertically and horizontally, forming a mass instead of a single column. This necessitated designing two shapes for the cups, bowls and plates—one concave and one convex.

Another innovative element was the complex color scheme, using four main color families each consisting of twelve harmonious tints and shades. This allowed the consumer more flexibility in color choice.

The concept of interrelated parts was extended to include coordinated glassware and stainless steel flatware.

Plaster molds reflect final form decision.

The Transition collection evolves from a square to a round shape.

Profile illustrates fidelity of shapes.

Designer Gerald Gulotta
Manufacturer Block China Corp.

Lauffer Magnum Stainless Steel Flatware 1968

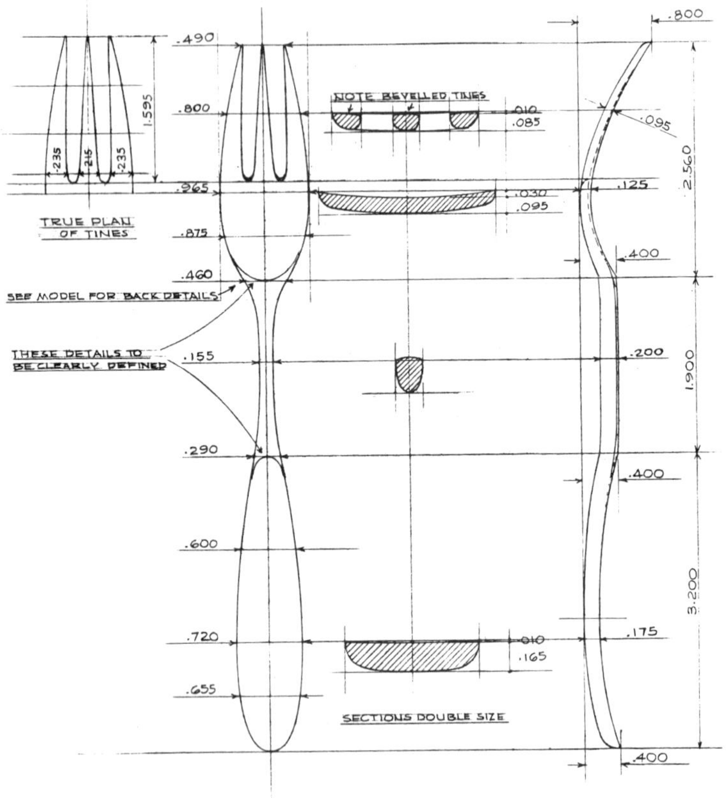

Metal models and precision engineering drawings are sent to the tool and die maker, and Wallance always checks the first samples.

The Design Problem: To design high quality plastic flatware for long term use.

Don Wallance had designed stainless flatware for Lauffer for a period of thirty years. By the late 70s, he felt that plastic material and technology had reached a stage of production capability and that a market existed for the manufacture of high quality, durable plastic flatware. It took several years for Wallance to resolve the doubts of the Lauffer Company, for whom this new direction involved the risks of investment and tooling costs. Through extensive technical research, Wallance determined that LEXAN polycarbonate was the high performance plastic that best met the objectives of the project with respect to physical, thermal, chemical, appearance and molding properties.

Wallance's goal was to design a product whose appearance, scale, functional and handling qualities would distinguish it from cheaper disposable flatware. The models illustrate his movement away from designs characteristic of stainless flatware to the simple, sturdy form appropriate to plastic.

Don Wallance works in three dimensions, experimenting in wood, and presents initial concepts in model form; the fork and knife are designed first; when these are approved, he proceeds with the other place setting pieces.

The metal models, formed from a lead and tin alloy, closely resemble stainless.

Sequence of production models from steel bar to finished spoon. Forms are implemented from blocks of steel. While most stainless flatware is made from sheets, Magnum requires an expensive forging process; the skill and care required for the final griding, polishing and articulation of detail add costs.

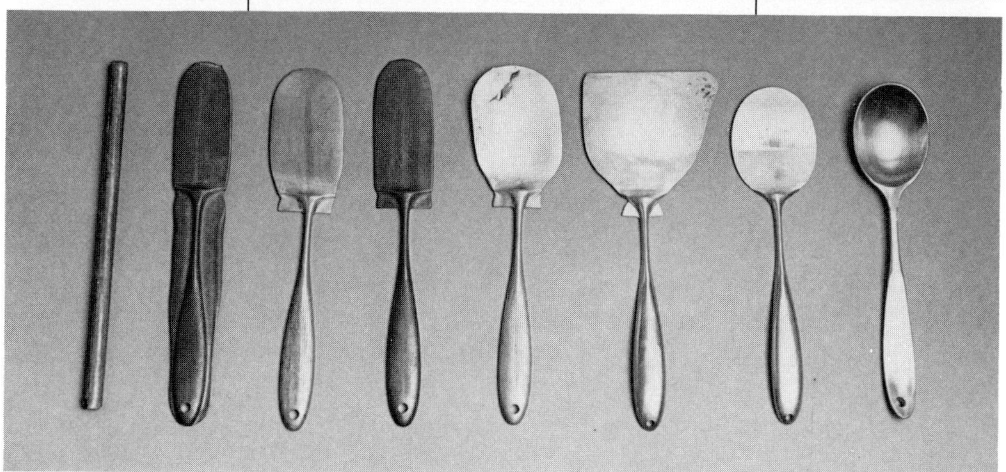

Designer Don Wallance
Manufacturer Lauffer (A Towle Company)

Block China Chromatics China, Glassware and Flatware 1971

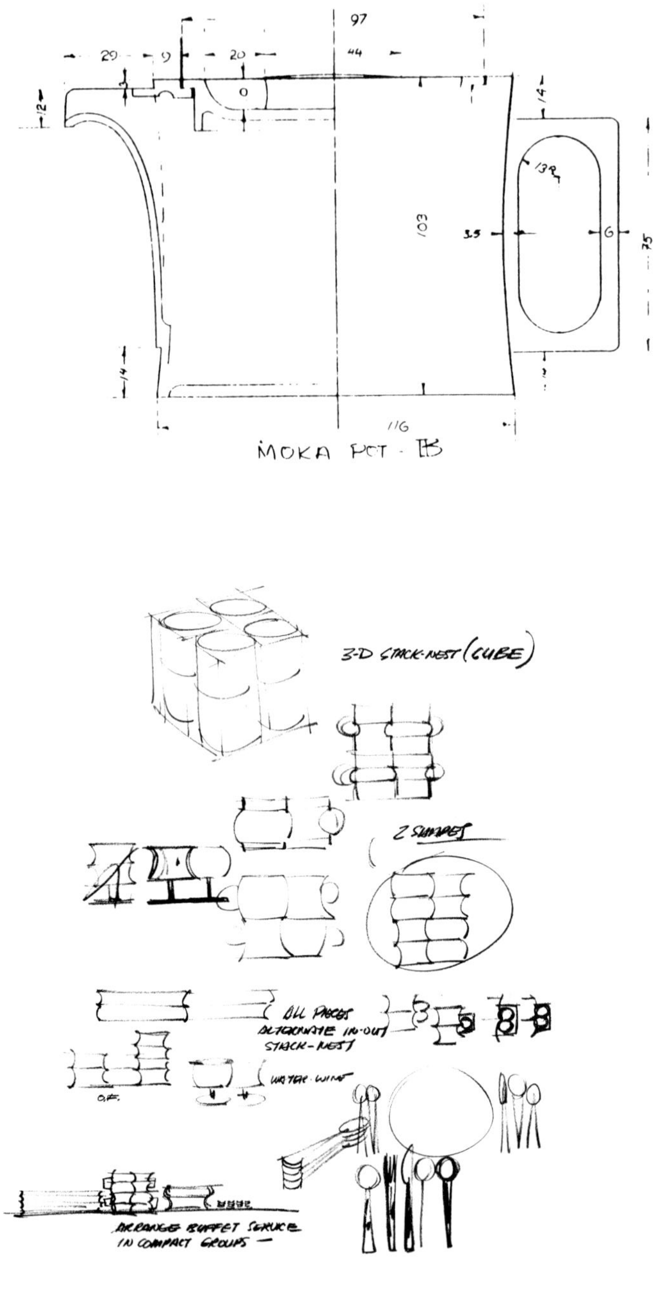

Concept sketches by Gerald Gulotta.

The Design Problem: To develop a contemporary porcelain dinnerware form that would make a continuing modern statement in a classic manner.

The entire dinnerware shape evolved from the idea that a square shape would be a pleasing form. Its success is based upon the interplay of shapes. These shapes, however, had to be mass produced on machinery that normally makes round shapes. From this constraint came the idea of "transition" from square to round. The tea cup and saucer best illustrate the transition: the square, forming the rim of the saucer, makes a transition into a circle at the rim of the cup.

During a three-month testing period, many samples were produced, carefully supervised by the designer, before the mass production stage was reached.

The evolution of one idea into another and a combination of practical considerations and intuition directed the process to its final conclusion. The designer was also fortunate in having an imaginative client.

Gulotta states that "the design of any ceramic product is for me a means of personal expression and enjoyment, especially as it allows me to work three dimensionally. These are cultural artifacts which can express artistry as well as function. They are created in a time frame and may remain there—or they can live forever."

The inter-related forms, with concave and convex shapes, stack vertically and horizontally, and relate to coordinated glassware and stainless steel flatware.

Designer Gerald Gulotta
Color concepts Jack Prince
Manufacturer Block China Corp.

★ Copco Cast Iron Cookware 1969

Drawing of teapot and top design by Michael Lax.

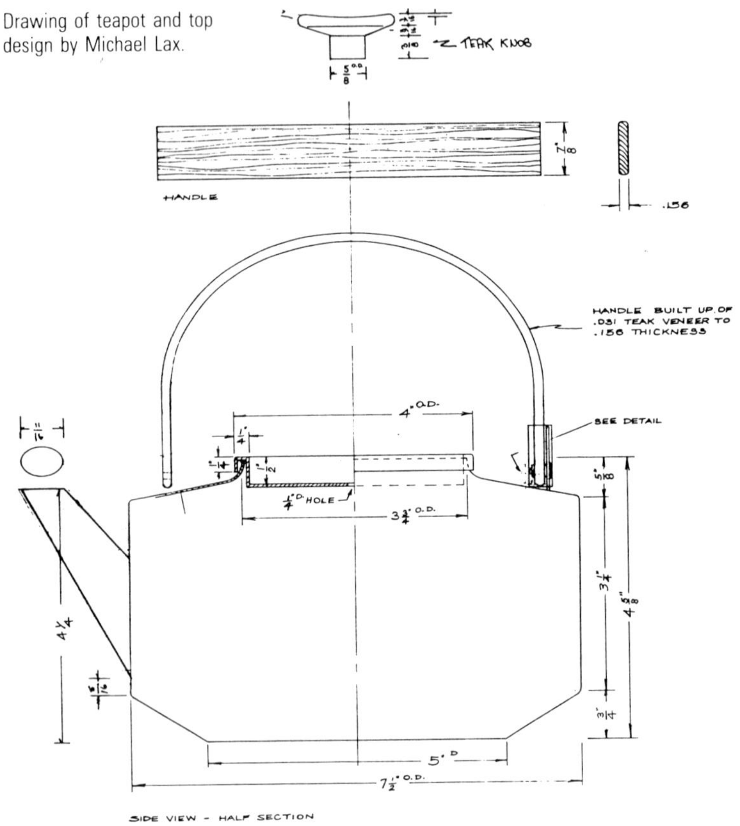

The Design Problem: To develop a series of products that are excellent cooking vessels and esthetically appealing for table service.

Early research was done by consumer interviews to identify specific cookware items to be designed. The new consumer interest in international foods spurred the design of the *paella* pan.

Cast iron was an accepted cooking material; adding the porcelain enamel coating in bright colors enhanced its appeal, as provided protection against rusting and made cleaning easier. This stage in the development required extensive experimentation by factory engineers to create new casting techniques.

The familiar Copco teapot has classic proportions.

Cast iron oven platter has a graceful look that disguises its weight.

Designer Michael Lax
Manufacturer Copco, Inc.

Heller Stacking Mugs 1972

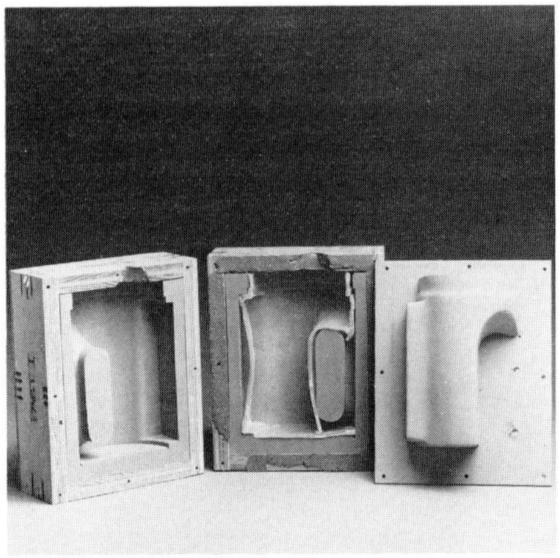

Mold shows the new closed gap at the top of the original cup handle.

The Design Problem: To develop a mug from an earlier cup design.

As the result of research into consumer preferences, the designer decided to close the gap at the top of the original cup handle. The new handle design forms a continuation of the mug wall and is a totally innovative concept. A pitcher incorporating this same handle was designed later.

The stacking mugs, which have become contemporary classics, reflect the Vignelli approach to design: "discipline appropriateness, ambiguity."

Heller stackingware is in the Design Collection of the Museum of Modern Art.

The pitcher incorporating the same form was designed later.

The stacking mugs and platters have become contemporary classics and reflect the Vignelli approach to design.

Designer Massimo Vignelli
Manufacturer Heller Designs, Inc.

Flatware and Dinnerware, AA In-Flight Service 1974

Close up shows concave shapes of spoon and fork tines.

The Design Problem: To create distinctive in-flight dinner service at the same or lower cost as previous service.

Cost considerations directed the design decision. The original design, which would have required forging of all flatware pieces, was modified so that the fork and spoon could be stamped. The originally planned six-piece setting was reduced to three. Surfaces are flat so that pieces fit together compactly. The in-flight service is characterized by a classic simplicity of form and by the carefully balanced proportions of the flatware.

The surfaces are flat so that pieces fit together compactly.

The in-flight service is characterized by a classic simplicity and balanced proportions.

Design consultants Henry Dreyfuss Associates
Supervision (partner) Neils Diffrient
Designer (associate) James Ryan
Client American Airlines

Peppermate 1982

The Design Problem: To design a peppermill that would appeal to the serious cook, could be used at the table and would collect measured amounts of pepper for cooking.

From the beginning, the most important consideration was to design a peppermill that worked well. Research showed that existing products were poor mechanically and were difficult and uncomfortable to use. In the Peppermate's design, the shape of the base and the large turn key contribute to smooth operation and eliminate fatigue for the serious cook. In the removable tray, an idea generated by clients Ken and Patti Ferren, pepper can be collected and premeasured.

The designer's belief that form follows function is expressed in the clean lines and timeless appearance of the product.

Engineering drawing.

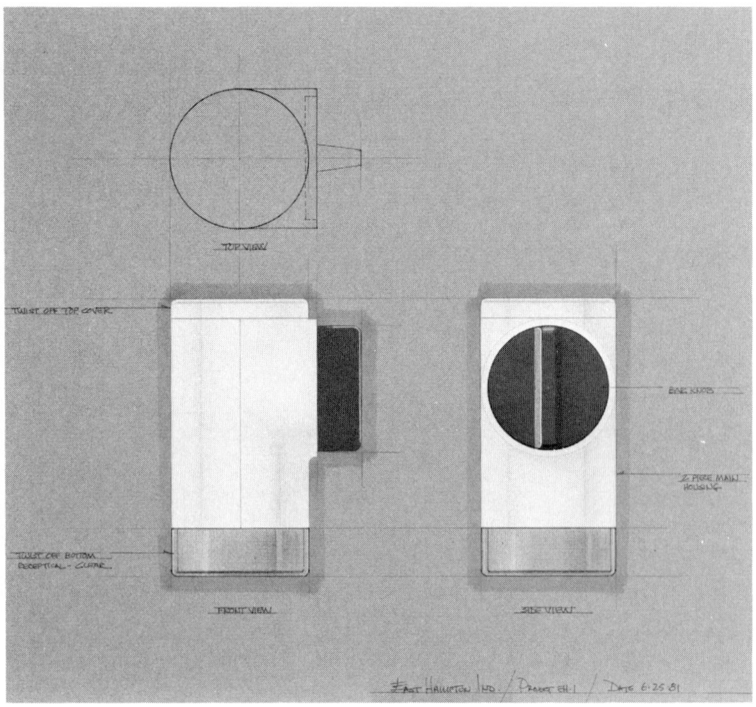

Open and closed views of a peppermill whose mechanical action and leverage permit smooth grinding.

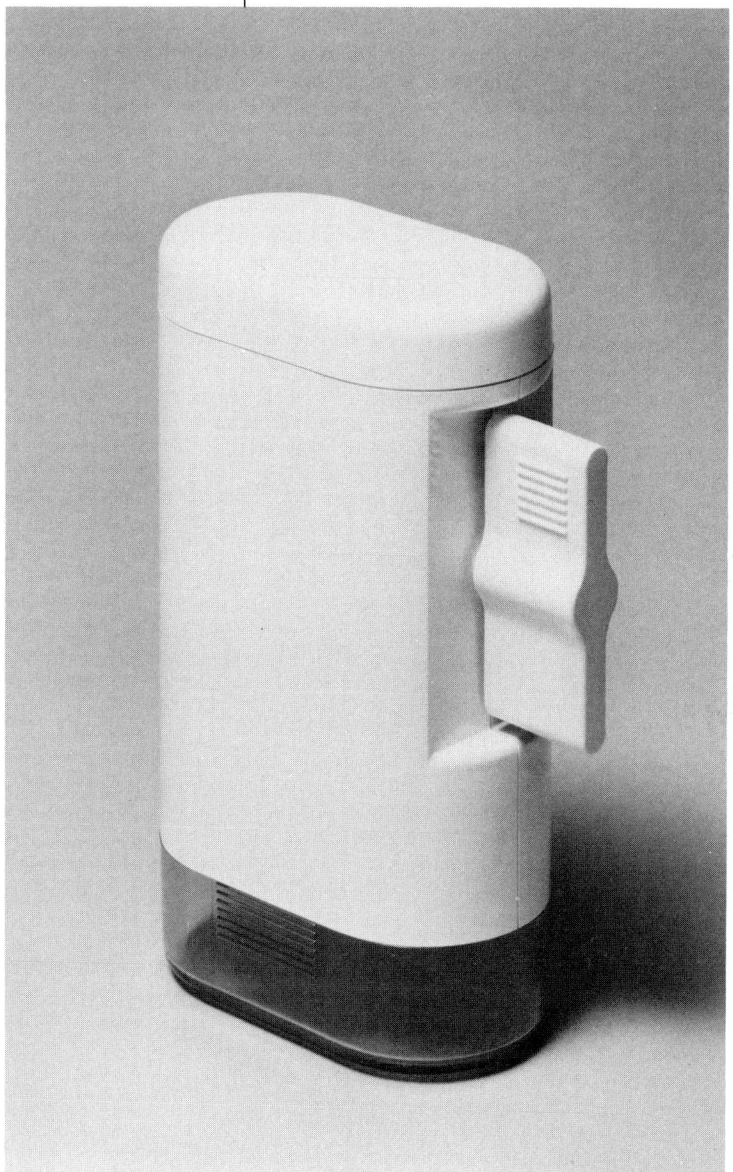

Design consultant Ronald Emmerling Design, Inc.
Designer Ronald Emmerling
Manufacturer East Hampton Industries, Inc.

★ **Maxim 4 Convection Oven** 1980

Early concept sketches reveal a search for form.

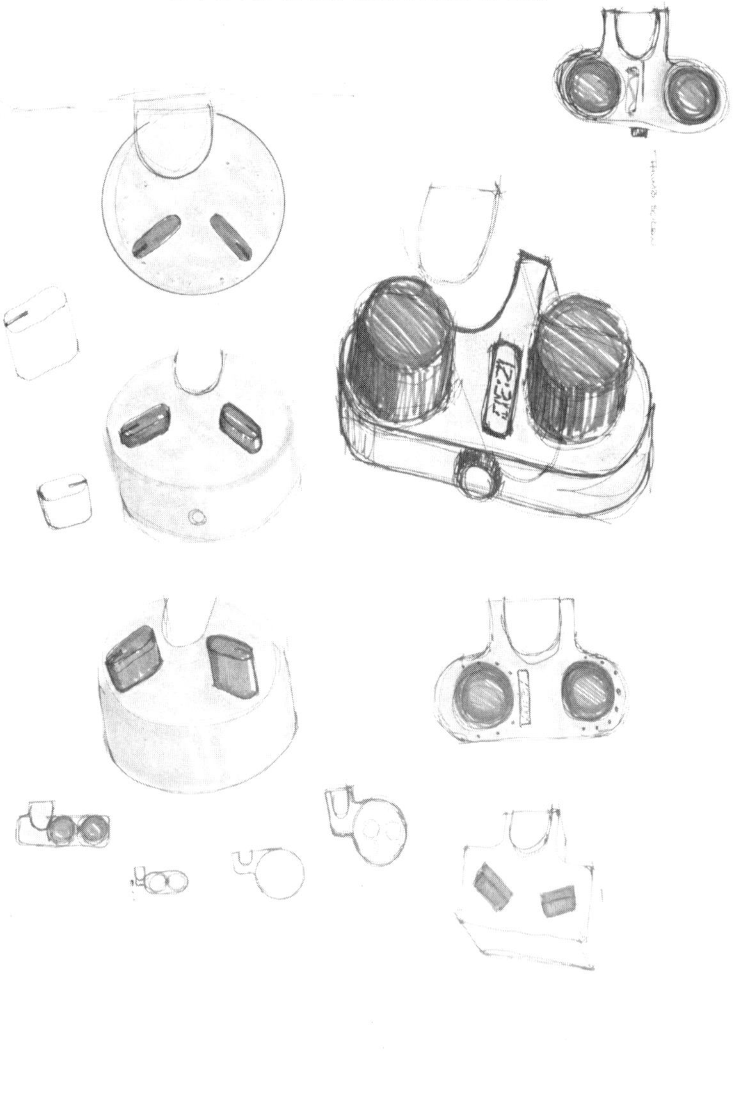

The Design Problem: To redesign the convection oven, usually found in a professional kitchen, for use in the home.

This project evolved from the combined needs of production, function, marketing and aesthetics.

Early concept drawings indicate a search for form: the control module and legs.

The designer, Morison Cousins, states that "at the time we designed this oven, we were striving to bring a sense of warmth to our products, to make them more 'personable.' The product is a fine utensil. And so, we balanced the serious (brushed chrome and bold black forms) with red details, and oversized, pad-like feet. Serious products can be fun!"

The Cousins' design approach shows traces of European influences: the precise and disciplined German design, Scandinavian design of the 50s and early 60s with its crafted sense of human involvement and the wit and totally human approach of Italian design.

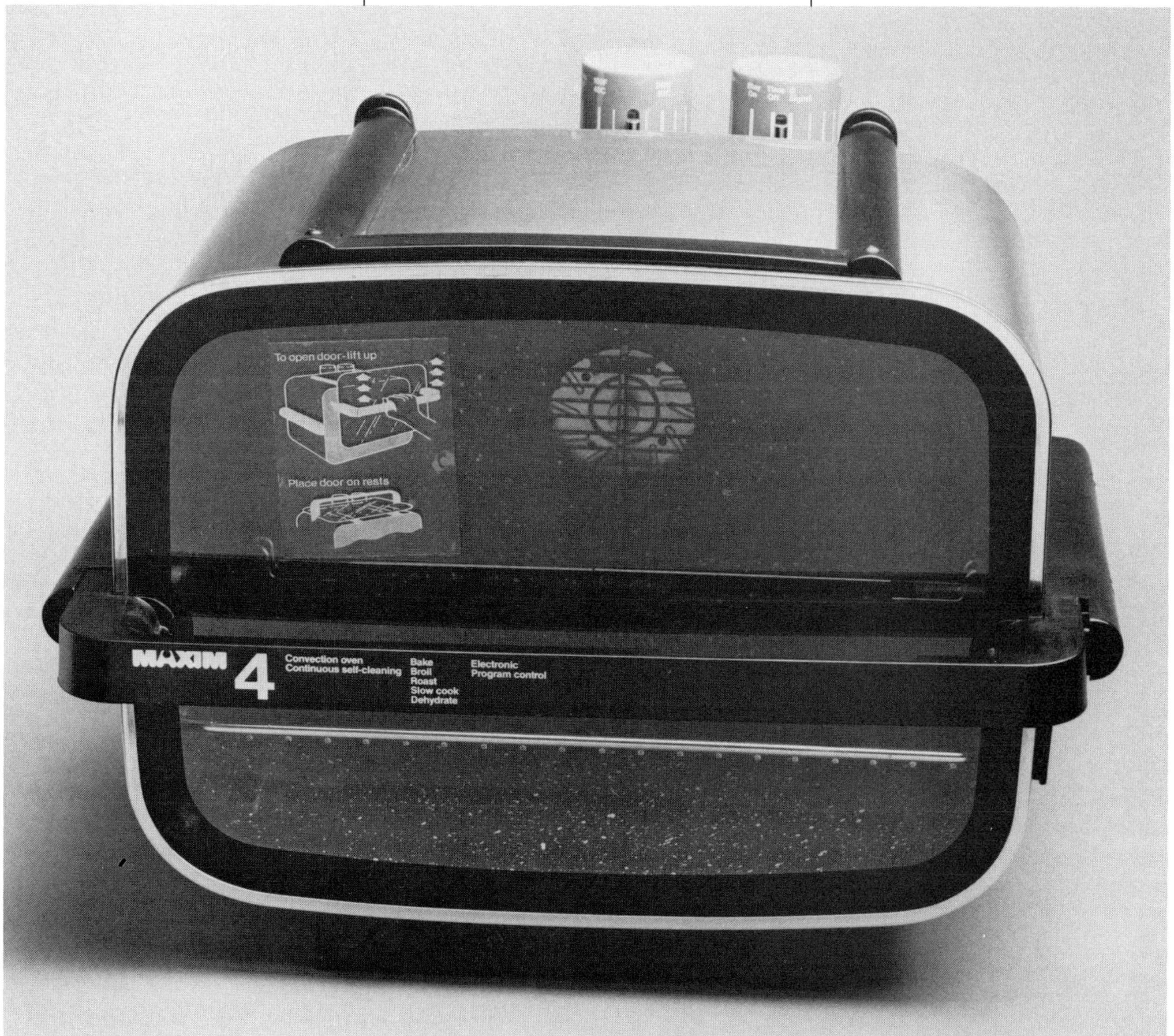

Design consultants Morison S. Cousins + Associates
President and design director Morison S. Cousins
Designer John Lonczak
Manufacturer The Maxim Company
President and engineer Joseph Bender

The convection oven, usually found in professional kitchens, was redesigned for private homes.

Tupperware Modular Mates™ Containers 1982

Research indicated consumers wanted transparent containers in a range of sizes.

The Design Problem: To design space saving food storage containers.

Research was conducted on food storage practices in major countries of the world. Results indicated consumers wanted a transparent container and a range of sizes.

The design staff developed a modular configuration, eventually selecting 16 container shapes, which fit together for space efficiency. These containers were designed to be "workhorses" of kitchen food storage.

Developed in a freezer-safe material, each unit has a see-through area. An alignment feature was added to the container's cover to guarantee perfect fit.

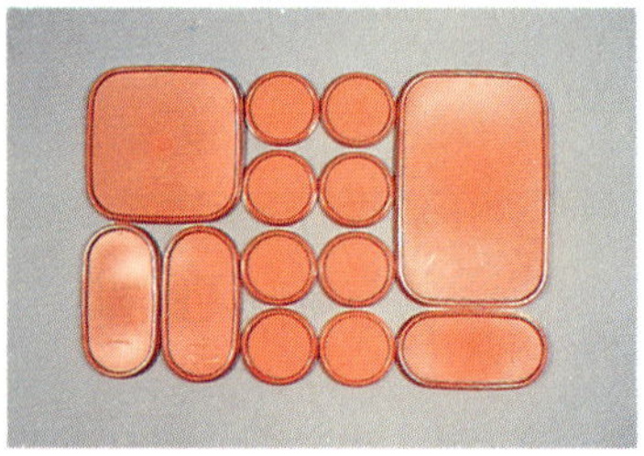

Sixteen container shapes were designed which fit together to save space, and each has a clear see-through area.

Manufacturer Tupperware Company
Design Product development department staff

Eureka Mighty Mite Vacuum Cleaner 1982

The Design Problem: To develop a compact, easily portable vacuum cleaner with full suction capabilities, for use in residential situations.

Research for the Mighty Mite focused on a variety of areas. Determination of unit size and bag capacity as it related to the form and proportions was important. Following this phase of research, which was carried out in wood and foam core models, the designers felt that a narrower, taller shape was desired over a wider, flatter shape.

The basic design problem was solved when, in the third wood model, it appeared that form and function had combined to produce a vacuum that worked as intended: a compact, full suction, highly mobile cleaner, that also appeared compact, powerful and mobile.

The Mighty Mite is composed of simple design elements with bold detailing in the switch, latch and wheels. The overall shape is straightforward and functional in appearance, yet contains a degree of softness.

The industrial design department was able to contribute a major share to the total design of the Mighty Mite because, as they worked out functional and esthetic details, they also considered basic engineering, manufacturing requirements, and marketing needs.

The shape and size evolved through models, and the narrower, taller shape was selected over a wider flatter form.

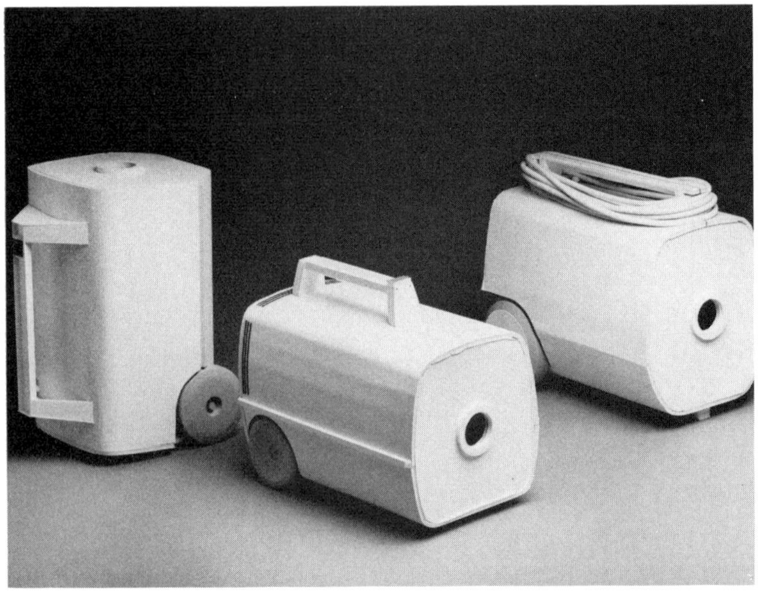

The objective of the final design was to signal a highly mobile vacuum cleaner that was compact, easy to carry and store, and yet still powerful.

Manufacturer The Eureka Company
Chief industrial designer Samuel E. Hohulin
Industrial designer Kenneth Parker
Project engineer William E. Hanold
Design draftsman Franklin C. Phifer
Chief engineer Willis E. Berry
Manager of engineering Bill J. Hart
Vice president technical operations Richard E. Kronmiller

Gillette Pro Max Compact Hair Dryer 1975

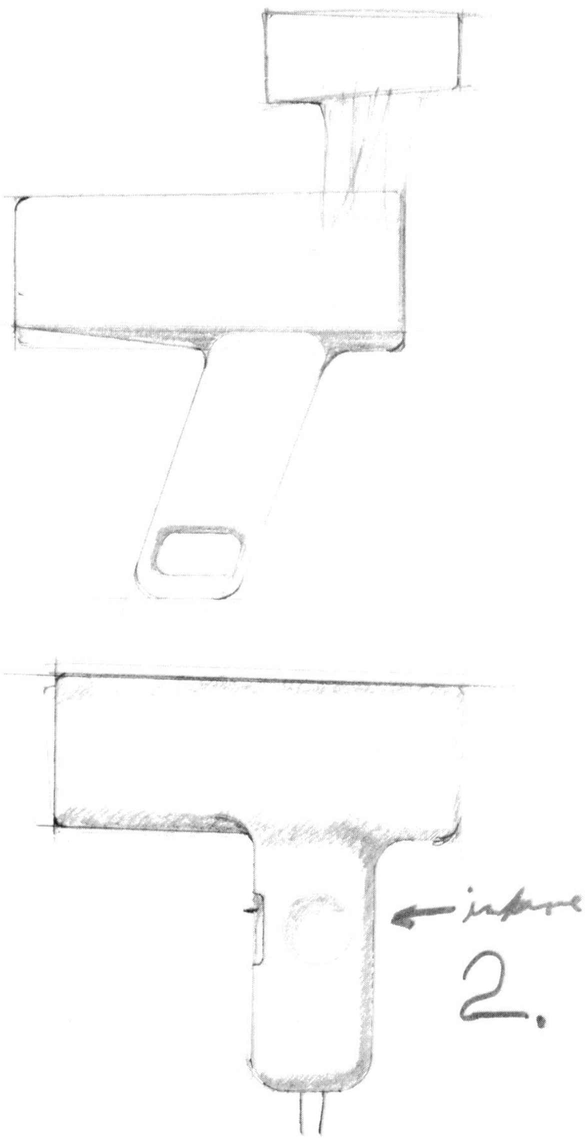

concept sketches.

The Design Problem: Using defined motor, fan and heater, to develop a form appropriate for a hand-held hair dryer which would be marketed by a major company seeking a substantial share of the market.

As part of the research, the designers examined similar models in the marketplace. Various concepts were developed.

"The moment of enlightenment occurred when, after exploring substantial numbers of forms, we agreed that the extremely simple form, consisting of two cylinders, was the most aesthetically satisfying and comfortable to hold and use." The product relies on basic forms and careful detailing to achieve its character.

The design is part of the Design Collection at the Museum of Modern Art.

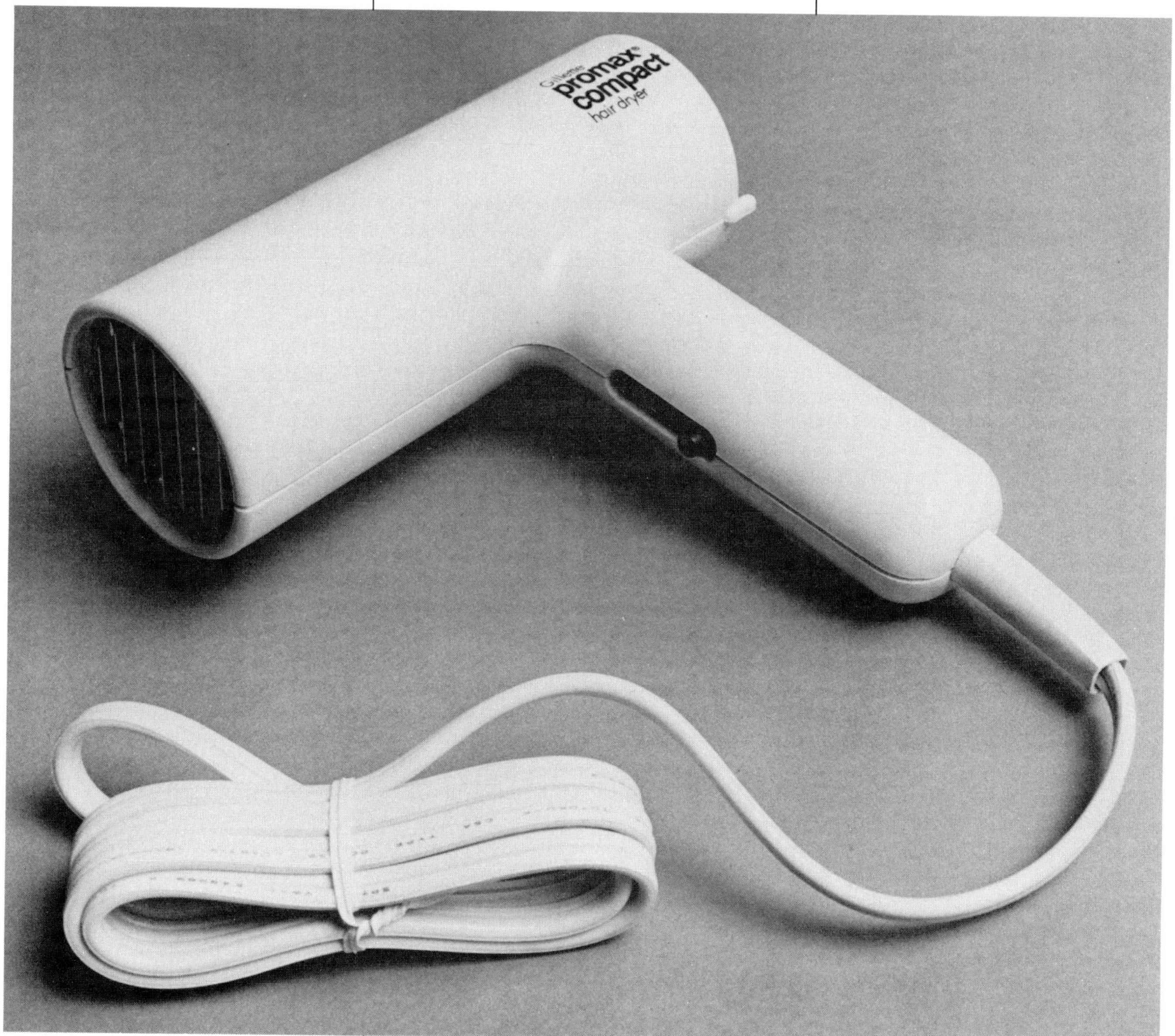

Consulting designers Morison S. Cousins + Associates
President designer Morison S. Cousins
Vice president designer Michael A. Cousins
Manufacturer Gillette Company

The hair dryer relies on basic forms and careful detailing to achieve its character.

Brandir Ribbon Bicycle Rack
1979

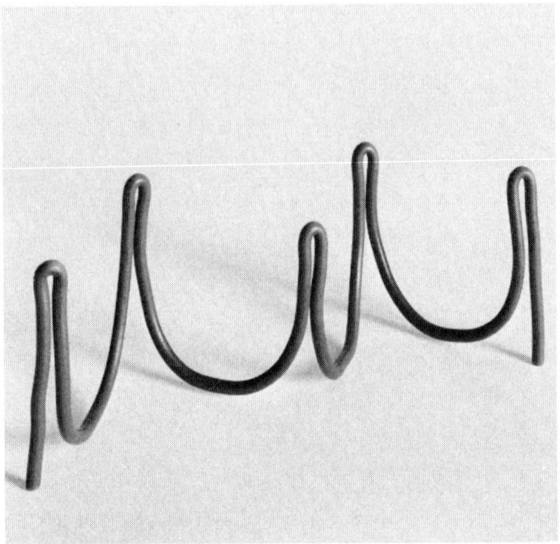

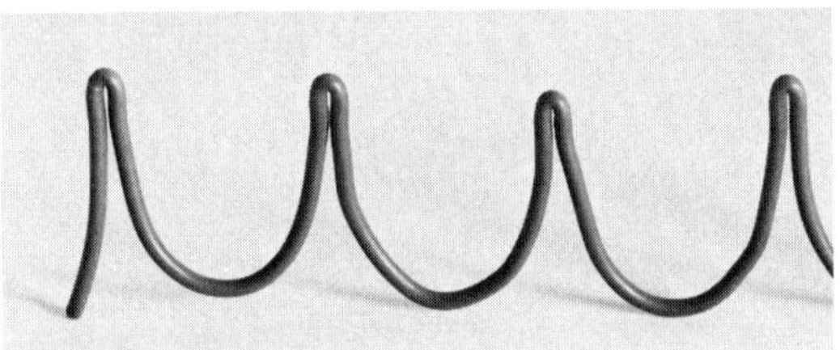

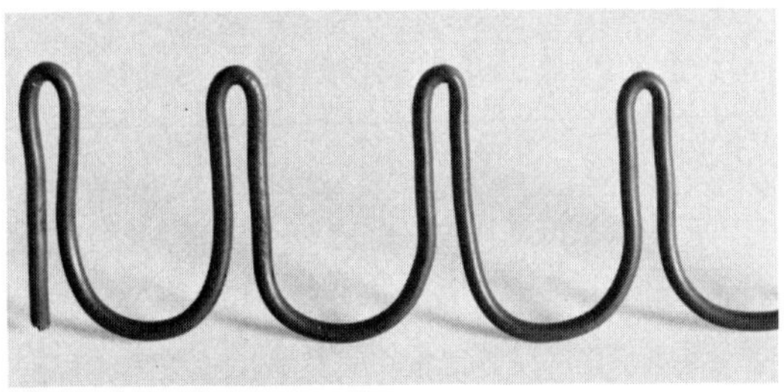

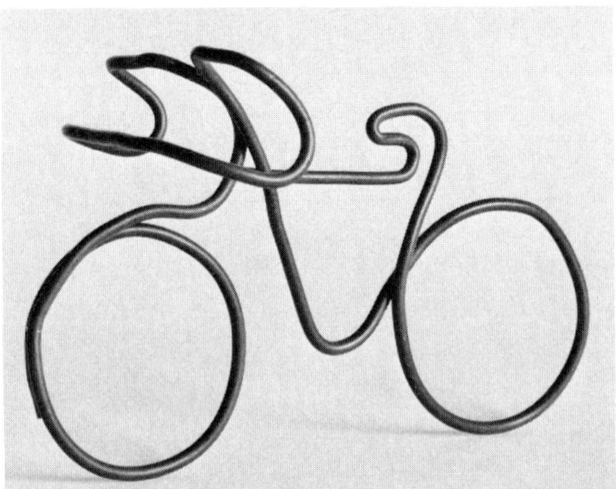

Steven Levine worked in wire to develop the basic form.

The Design Problem: To translate a sculpture into a working product design.

Steven Levine started his career as an industrial engineer and computer analyst. Working with doodles, and then wire, he formed a sculptural piece. As he refined this form, he discovered that it could have a useful purpose as a bicycle and moped parking device. Aesthetics were the main concern: how the size would relate to open spaces.

The rack is made of one piece of tubular construction (40 Steel Pipe) and can be mounted below grade. Levine is both designer and manufacturer of the Ribbon Rack. He has stated that "the form was an inspiration that found a function."

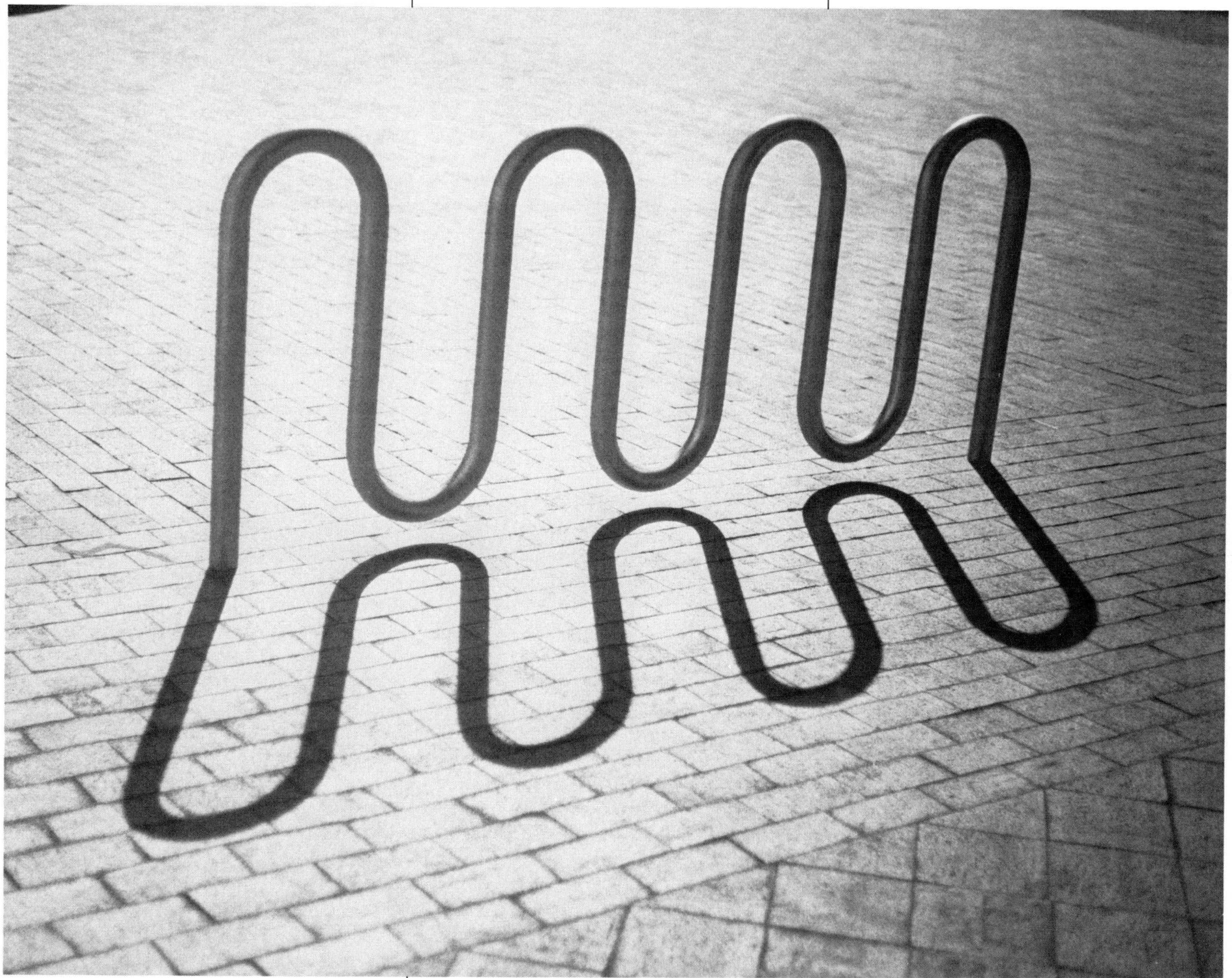

Designer Steven K. Levine
Manufacturer Brandir International, Inc.

Rack is made of one piece of tubular construction and can be mounted below grade.

Mattel Sunrunners Roller Skates 1979

The Design Problem: To design lightweight roller skates with a unique steering mechanism, simple assembly, quick response and full adjustability, at low cost.

Research indicated that a "low cut" running-shoe shape was adequate for recreational skates. Individual "trucks" (see drawings) attached to the steel chassis with rubber pads on either side vary the rebound rate and resistance.

Human factors and product safety were paramount considerations in the design. Simplicity and orientation to actual use characterized the design direction.

Exploded view of skate roller and toe steering mechanism.

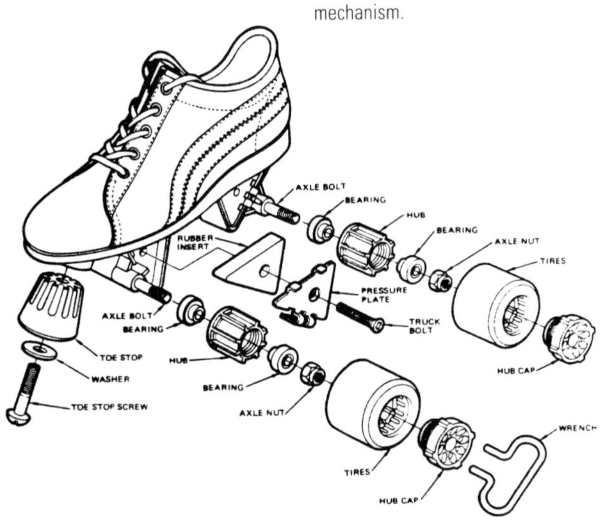

Design consultants The Stansbury Co.
President Benjamin H. Stansbury, Jr.
Engineer Brian Hills
Manufacturer Mattel, Inc.
Industrial designer Steve Kimmell

Lightweight skates built from low-cut running shoe shape permit a quick response and are fully adjustable.

Rezek Orbis Table Lamp 1982

The Design Problem: To develop a small unobtrusive, flexible task light that would reach over a desk area.

The idea of using a ball and socket as a universal pivot determined the form and direction of the product. According to Ron Rezek, his design "reflects my interest in reducing a form, even a functionally complex one, down to its essential three-dimensional shape."

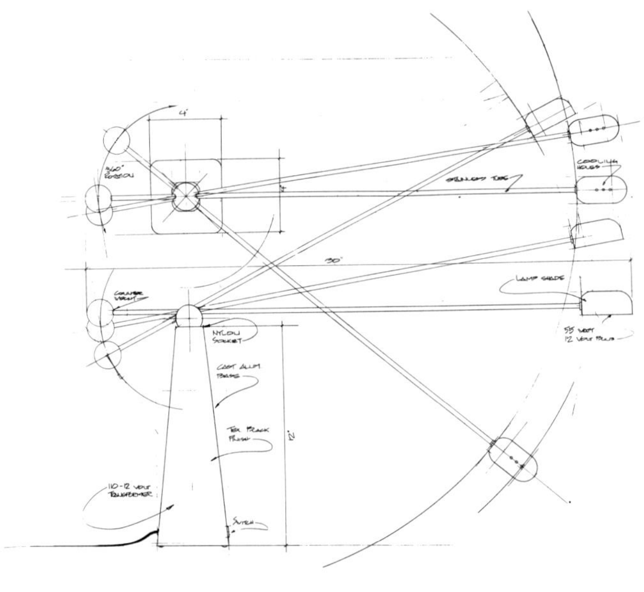

Detail drawing of arc and ball and socket design.

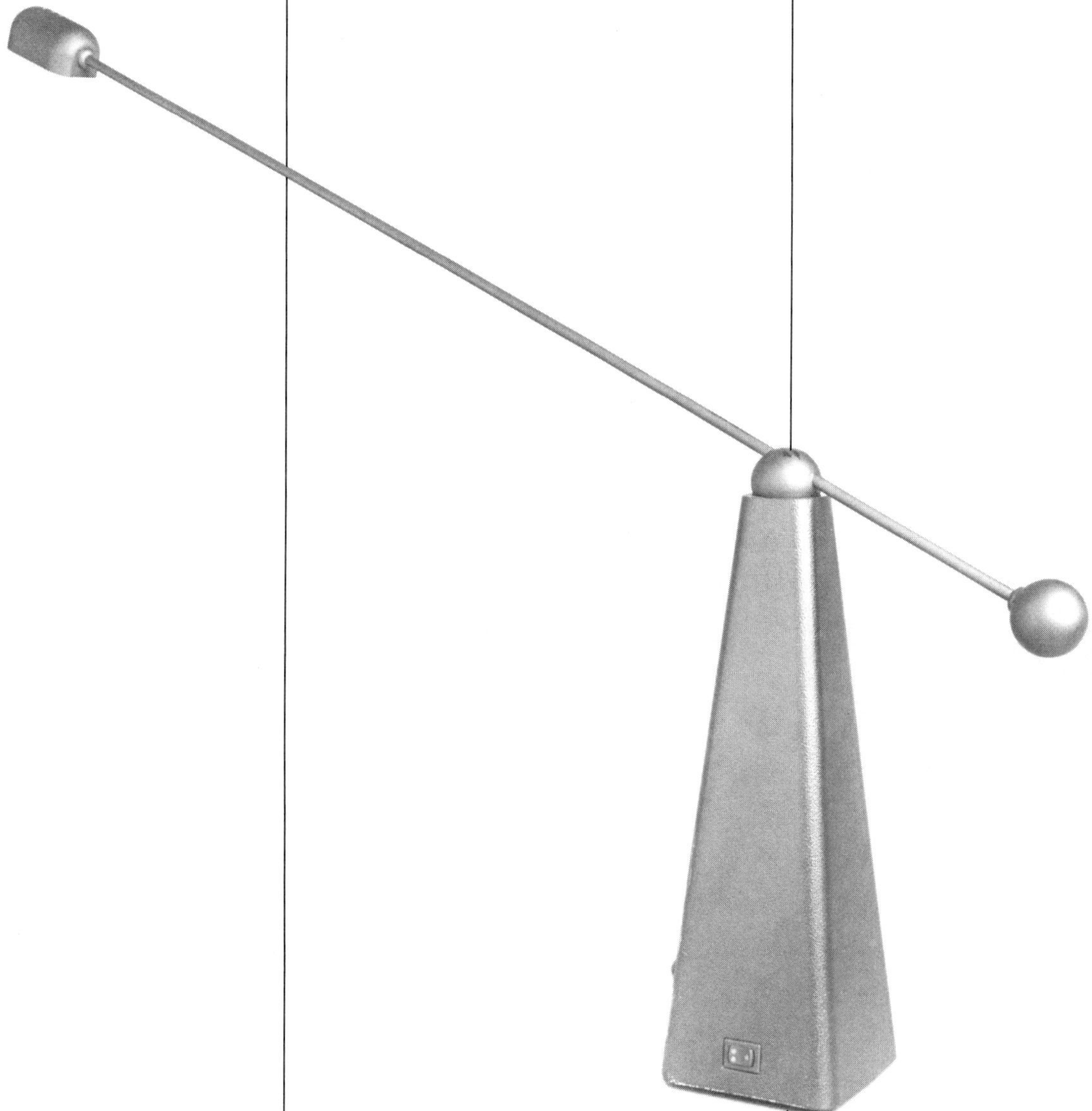

Manufacturer Ron Rezek Lighting & Furniture
Designer Ron Rezek

The aim was to develop a small, unobtrusive, flexible task lamp that would reach over a work space.

J.C. Penney MCS Series Stereo Receiver 1982

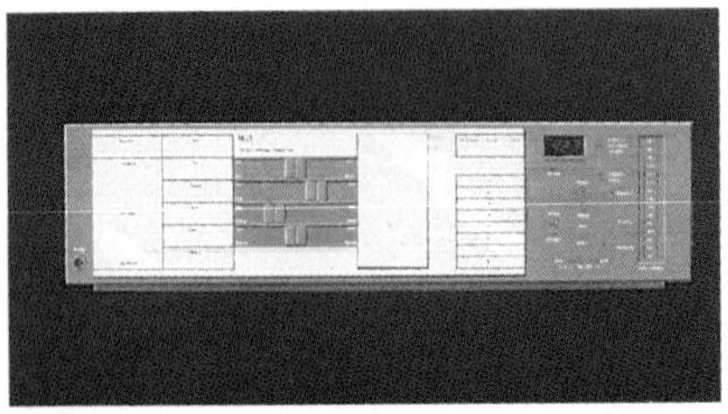

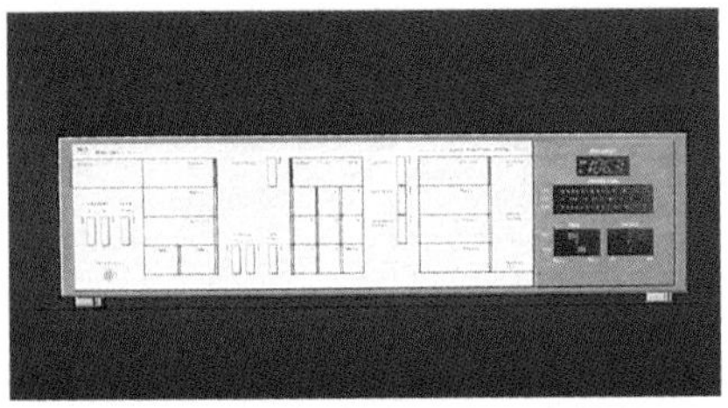

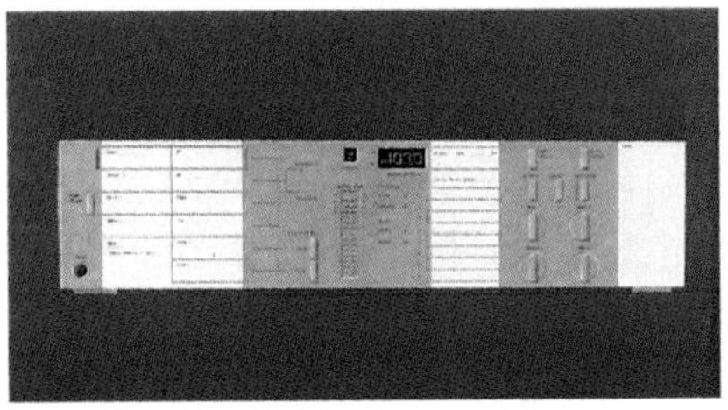

Color renderings of alternative solution for the front panel design.

The Design Problem: To simplify the operation of a highly sophisticated multifeatured product and to develop a unique functional appearance.

Task analysis research indicated that most receivers are intimidating in appearance and difficult to use. The designers decided to locate all controls in one dominant area, in a logical sequence with appropriate spacing. See early concepts (1) and (2). They established a vocabulary of three control sizes based on frequency of use. Additional convenient features include visual feedback through indicator lights, electronic displays and graphics.

Research of competitive products and available technology indicated the necessity of taking full advantage of microprocessor technology. Stock, off-the-shelf electronic components were used, reducing costs.

MCS Stereo and Receiver offer high quality performance and design at mass market price levels.

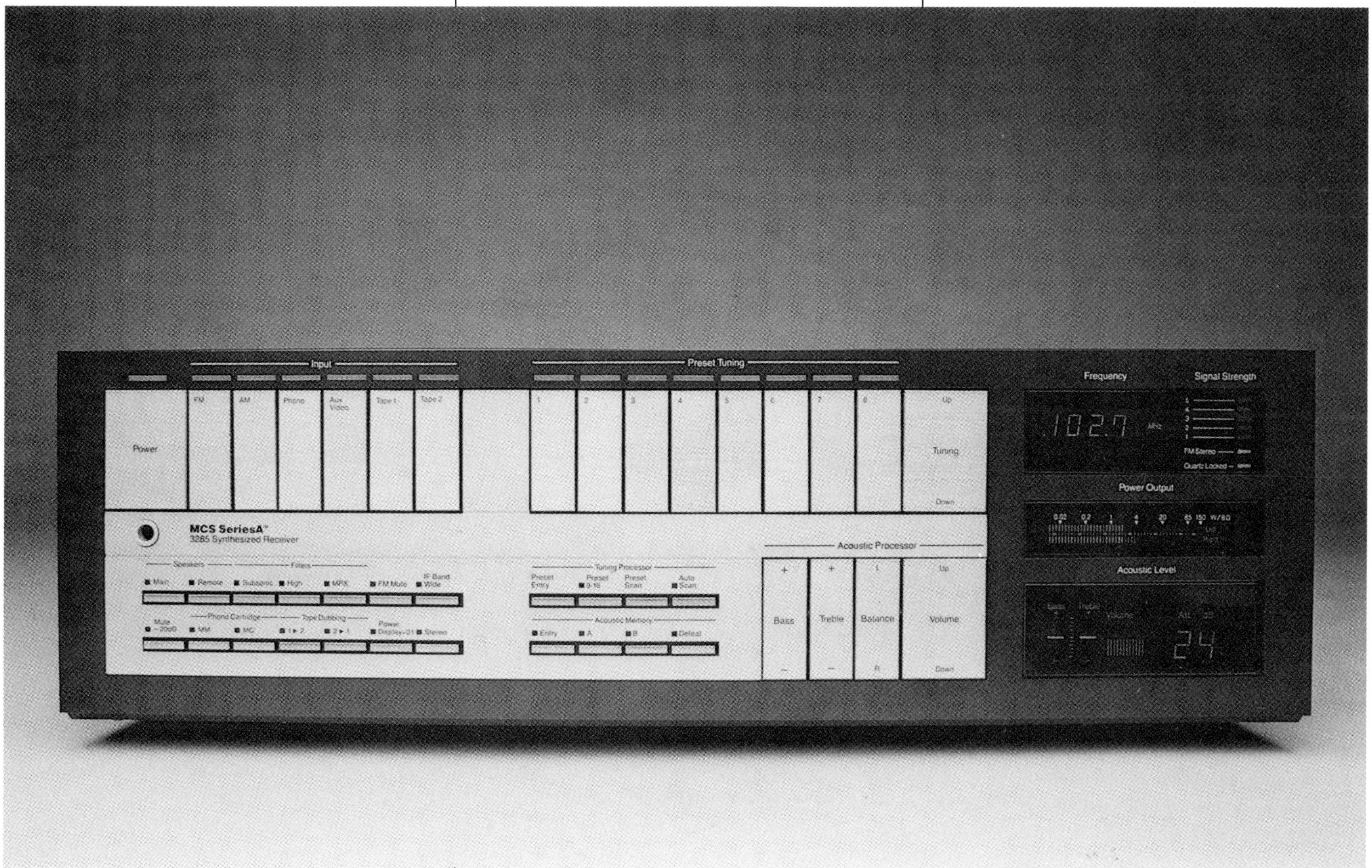

The receiver has a vocabulary of three control sizes based on frequency of use, part of its aim to simplify the operation of a sophisticated product.

Client J.C. Penney Co., Inc.
Project manager Jan E. Tribbey
Product designer Rick Blanchard
Manufacturer Matsushita Electric Industrial Co., Ltd.
Manager design and engineering S. Kimua

J.C. Penney MCS Linear Tracking Turntable 1982

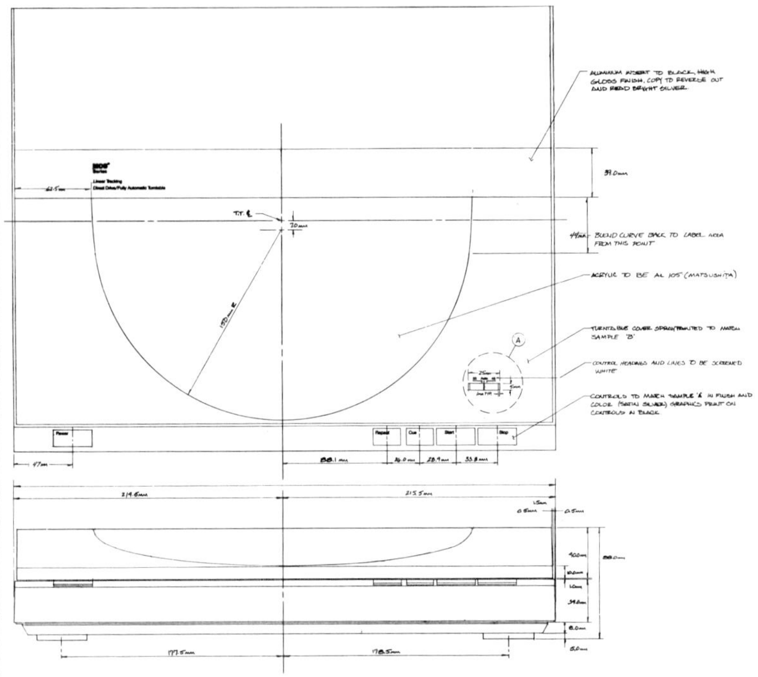

The Design Problem: To develop a product with a unique appearance that would maximize user operator convenience and appeal to a mass market.

The innovative use of a linear tracking mechanism mounted on the under side of the dust cover eliminates handling of the tone arm. Molding the dust cover in one piece of clear acrylic and selectively painting it reduced manufacturing costs and resulted in an elegant looking product.

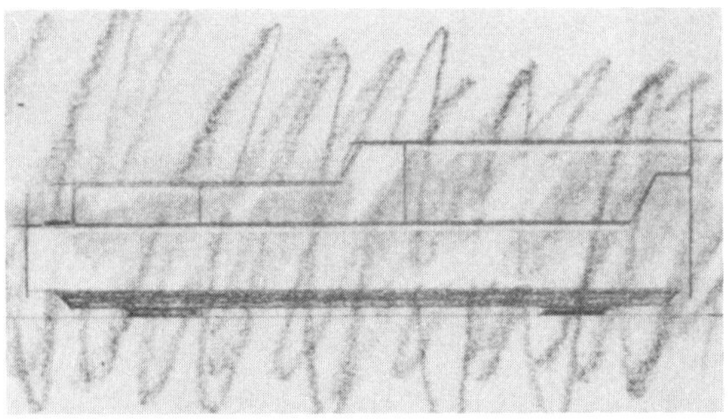

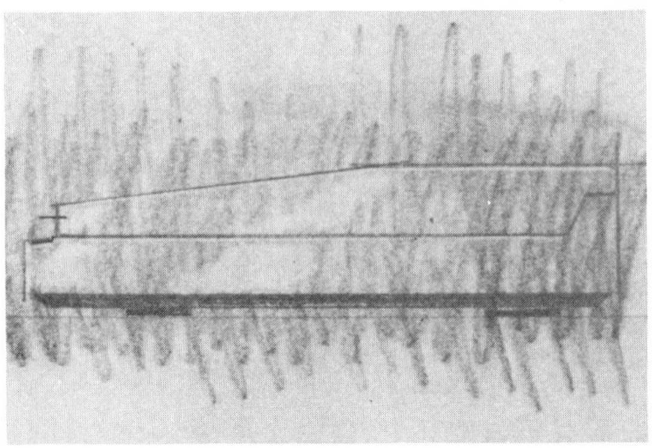

Presentation sketches reveal the development of the sleek profile.

The elegant and unique look of the JC Penney product proved that excellent design can appeal to the mass market.

Client J.C. Penney Co., Inc.
Project manager Jan E. Tribbey
Product designer Rick Blanchard
Manufacturer Matsushita Electric Industrial Co., Ltd.
Manager design and engineering S. Kimua

Zelco Pocket Fluorescent Lantern 1981

The Design Problem: To design the world's smallest battery operated fluorescent lantern.

Prior to this time, lanterns were used mainly by mechanics and sportsmen. This product was targeted for a broader market—one that would not equate size with value. A new electronic circuit made it possible to reduce the size. The unit is powered by four AA batteries.

The final design had to be esthetically pleasing to justify two requirements: higher price and acceptance of a new size and function.

The Zeller philosophy, "a clean design is an art form that invites handling and use" is embodied in this product.

The product is part of the Design Collection of the Museum of Modern Art.

Manufacturer Zelco Industries, Inc.
Concept Noel E. Zeller
Design Ben Zapolski, Fred Leung

The objective was to design the world's smallest battery operated fluorescent lantern and appeal to a broad, design-conscious market.

AMERICAN

4

MATERIALS

DESIGN

CLASSICS

Castelli Penelope Chair 1982

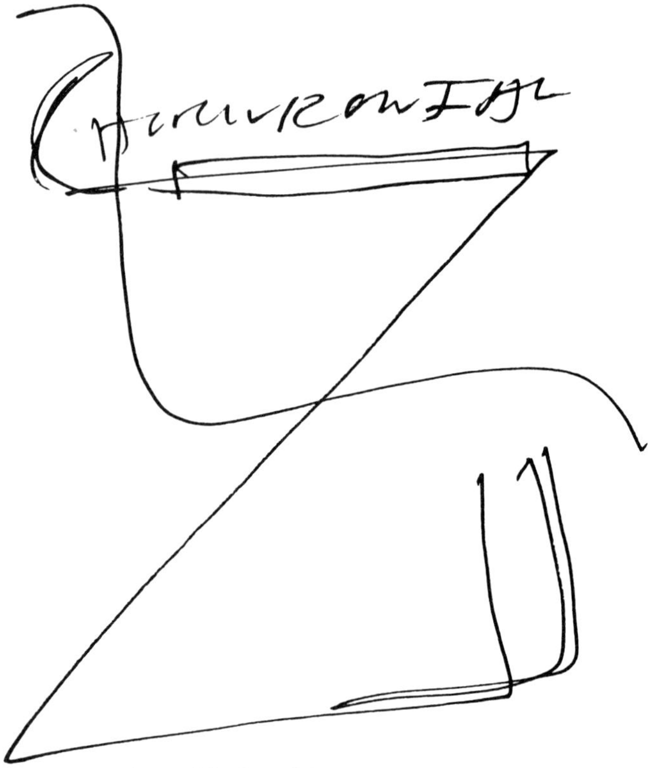

Charles Pollock's preliminary sketches.

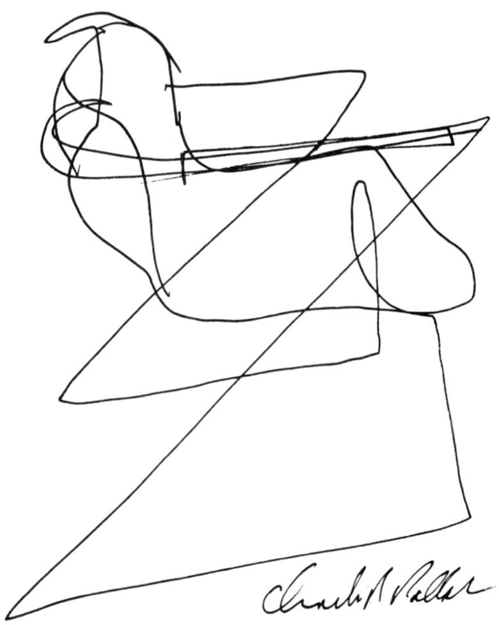

The Design Problem: To provide a small scale, mobile dining/stacking side-chair suitable for luxury interiors.

The frame of the Penelope is constructed of ½″ diameter chrome-plated NCD steel alloy rods with rubber arm rests. The seating shell is made of ¹⁄₃₂″ steel wire woven into a tight mesh, cut to size and pressure-molded to maintain contour. The combination of unusual materials, the thermoplastic resin finish applied to the woven steel and the 54,450 tiny air holes that comprise the seat shell blend to create a graceful and sturdy chair.

A graceful profile was achieved for this lightweight, portable, stackable and very comfortable chair.

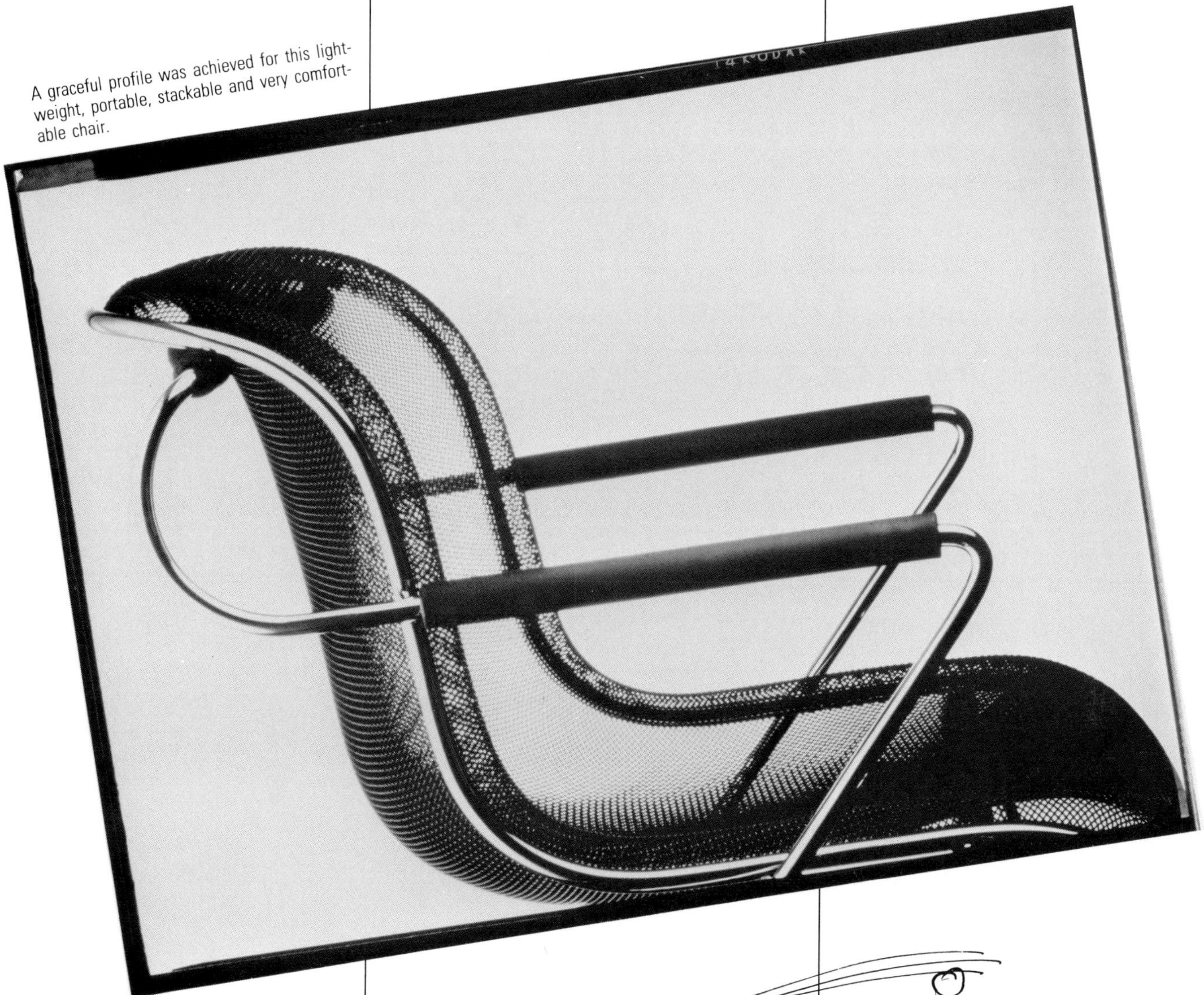

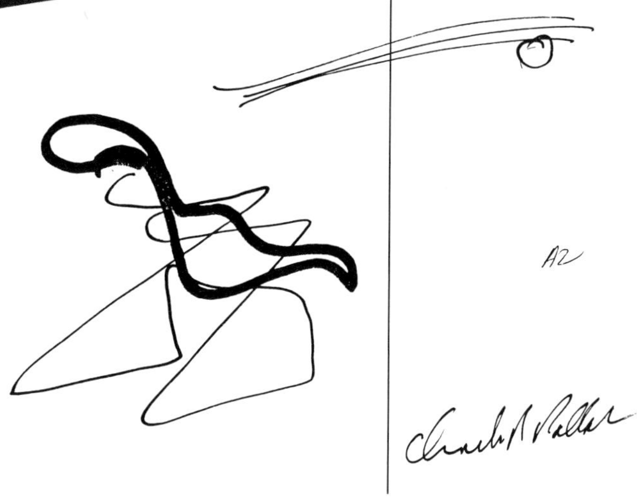

Designer Charles Pollock
Manufacturer Castelli Furniture, Inc.

Castelli Penelope Chair 1982

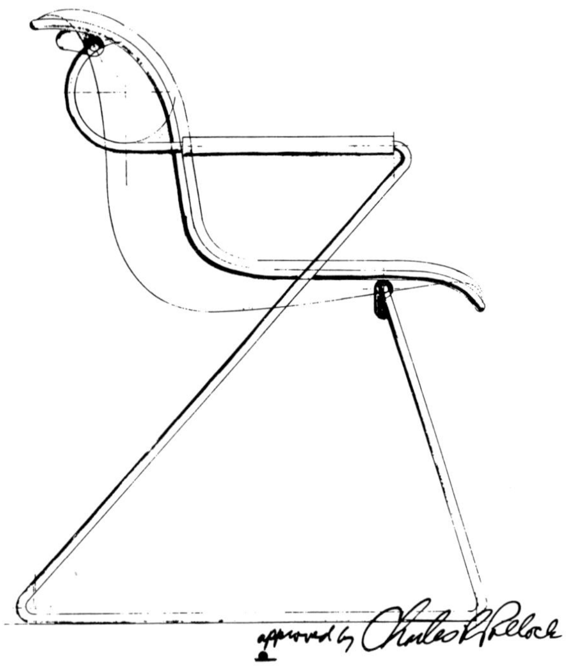

Charles Pollock's preliminary sketches.

The innovative use of lightweight material has produced a chair weighing only 20.6 lbs., one that is portable, stackable and extremely comfortable. Research into materials and structure was led by Castelli, Italy, in cooperation with the U. of Bologna's computer science department.

Thonet Sof-Tech Chair Collection 1979

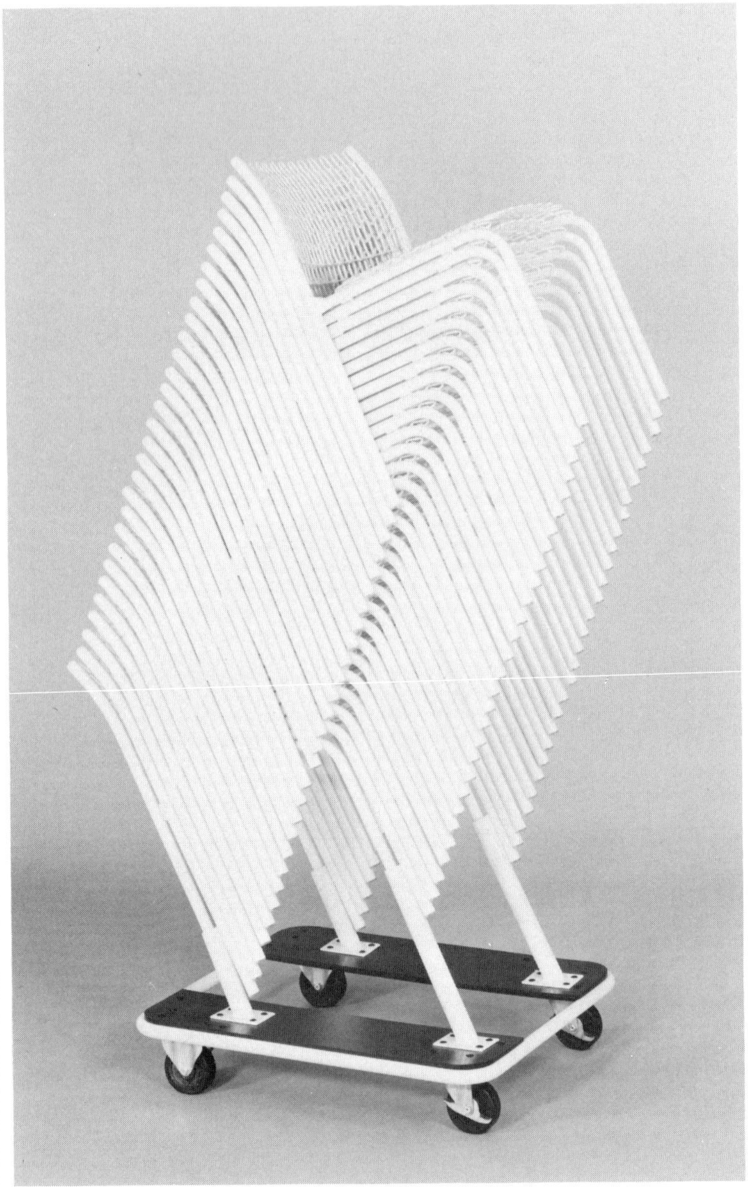

The Design Problem: To design a stackable dining or side chair that would be soft and pliable yet eliminate the need for reupholstery.

David Rowland first saw the unique spring wire in a workshop at Cranbrook Academy where he was a student, and he vowed to use this material in a chair design someday. Years later, this goal was realized when the designer created the innovative plastic coated spring wire chair. He had developed a process for dipping the properly aligned zig-zag springs into a rubbery plastic which was then cured with heat. The resulting SOFLEX material provides a soft seat that does not require upholstery for comfort.

Ergonomic studies were conducted using different spring gauges and coating thicknesses to determine optimum comfort. The flexible seat absorbs body heat and adjusts to each person. When the springs move, the PVC plastic moves at right angles to the springs. The frame has only one cross stretcher and has yieldable edges which permit the chair to move as the body moves.

Spring wire is usually hidden under upholstery. In the Sof Tech chair, the wire has become visible for the first time. David Rowland has achieved his goal of maximum comfort with minimum materials.

The goal was to develop a stackable side chair that would be soft and pliable yet eliminate the need for reupholstery.

Designer David Rowland
Associate Malcolm Peach
Manufacturer Thonet Industries, Div. Gulf + Western
Vice president design Joan Burgasser
Assistant Tom Biscoe

Samsonite Accord Attaché Case 1978

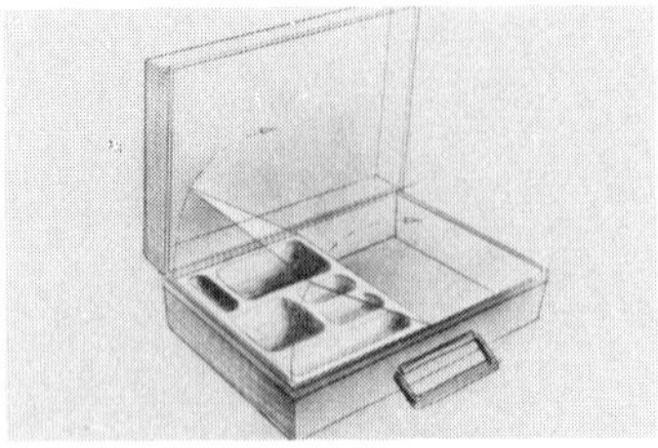

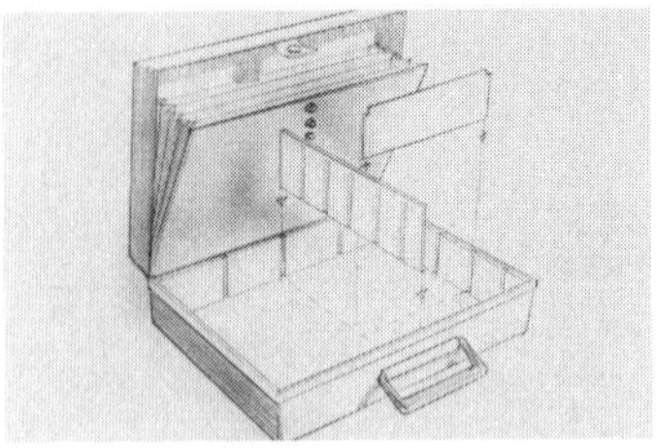

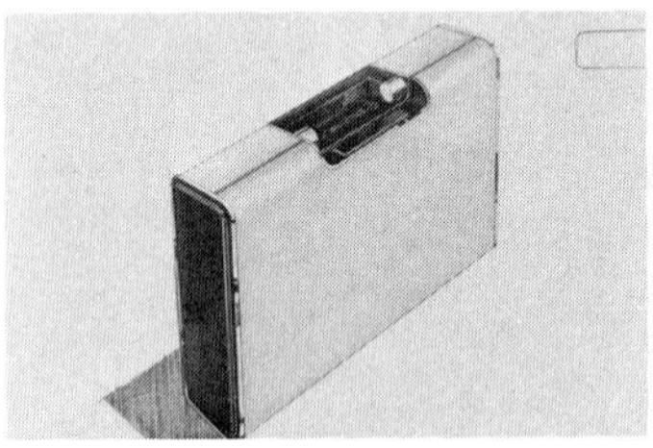

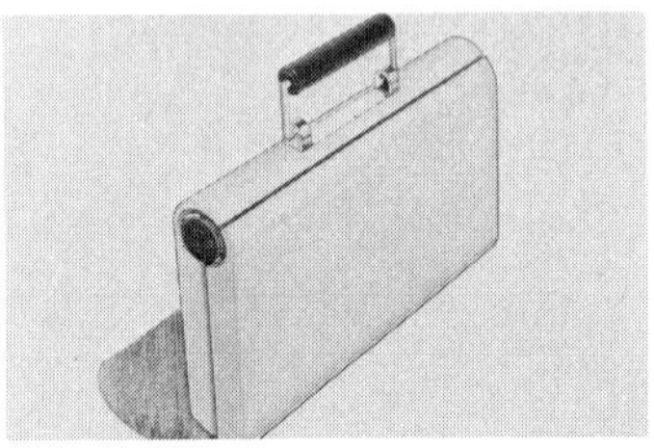

Concept sketches created for focus group research and to hone in on the final design.

The Design Problem: Using state of the art technology, to design an attaché case for mass market acceptance that would meet contemporary business needs, and have a quality image denoting good value at a competitive price.

In the Accord, one in the series of attaché cases that Design West has developed for Samsonite since 1962, the square look of previous models has been modified into a softer form for a more contemporary appearance. Extensive market research was conducted by the client to determine new consumer attitudes toward attaché cases and special functional requirements. These findings were incorporated into the design to satisfy a broad range of market segments.

In 1958, vacuum forming was a state of the art technology, which Samsonite brought into its factory in order to develop luggage that was economical and of a durability superior to anything on the market. In 1963, the original molded magnesium shell was replaced by ABS plastic for reasons of performance and cost. This material offers high impact and abrasion resistance and can be decorated with color.

Prior to the Accord, all Samsonite attaché cases had a die cast magnesium frame. On this project, the designers worked with engineering to develop a polycarbonate injection molded frame to reduce weight and increase durability. The client's engineering group then developed the technique for molding and bonding this frame to the basic shells of the attaché case.

Design consultants Design West
Design director Arthur E. Ellsworth
Staff support David E. Workman, J. Patrick Sullivan
Manufacturer Samsonite Corp.
Director of engineering Tom Booth
Staff support Marvin King, Larry Hill, Roger Winter

Prior to the Accord, all Samsonite attaché cases had die-cast magnesium frames, but to cut weight this model has an injection molded frame.

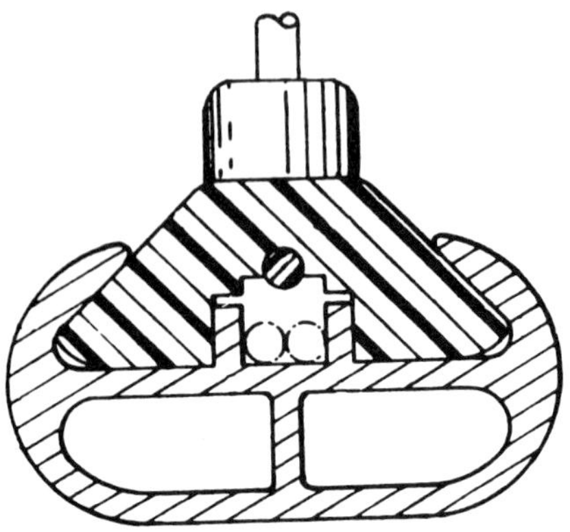

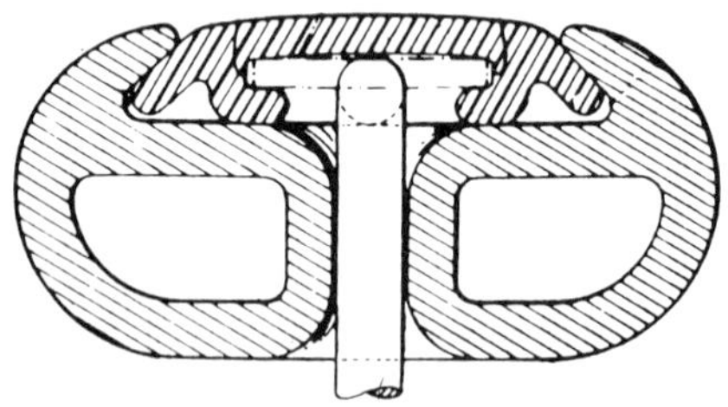

Aluminum Tennis Racquet 1968

The Design Problem: To design and engineer a professional quality aluminum racquet with the feel of wood and the power of metal.

The designers analyzed fiberglass, steel, wood, magnesium and aluminum and found that aluminum best met structural and manufacturing requirements. The state of the art of fiberglass was not yet sufficiently refined in 1967–70; graphite was still only utilized at the aerospace level and not economically viable; boron was still in the laboratory stage.

In developing the first aluminum racquet, the designers worked closely with Reynolds Aluminum. Technical analysis tests were performed with emphasis on the structural and functional components of the material. Latham & Brefka found that double hollow extrusion tubing would best meet the requirements of weight and could withstand the stress of drilling 52 holes (for strings) without collapsing the tube. This allowed enough room for a large support radius for the strings. At the same time Reynolds concluded that structurally the figure-8 extrusion was the answer and confirmed, through computer analysis, that this design passed longevity tests. Two weeks later they were player-testing the design solution.

Later, a more advanced racquet was manufactured by Seamco eliminating the holes in the frame and equalizing string tension.

Latham and Brefka's approach to design is "that the designer should be able to solve or at least understand and direct the technical solutions. We were solving the visual, functional, structural and production economics simultaneously."

Design consultants Latham & Brefka
Partner Peter A. Latham, Paul Brefka
Client Paul Sullivan Sports, Inc.
President Paul Sullivan

The first aluminum product opened the way for an array of new materials used for tennis racquets.

KBL Kryptonite-4 Bike Lock
1977

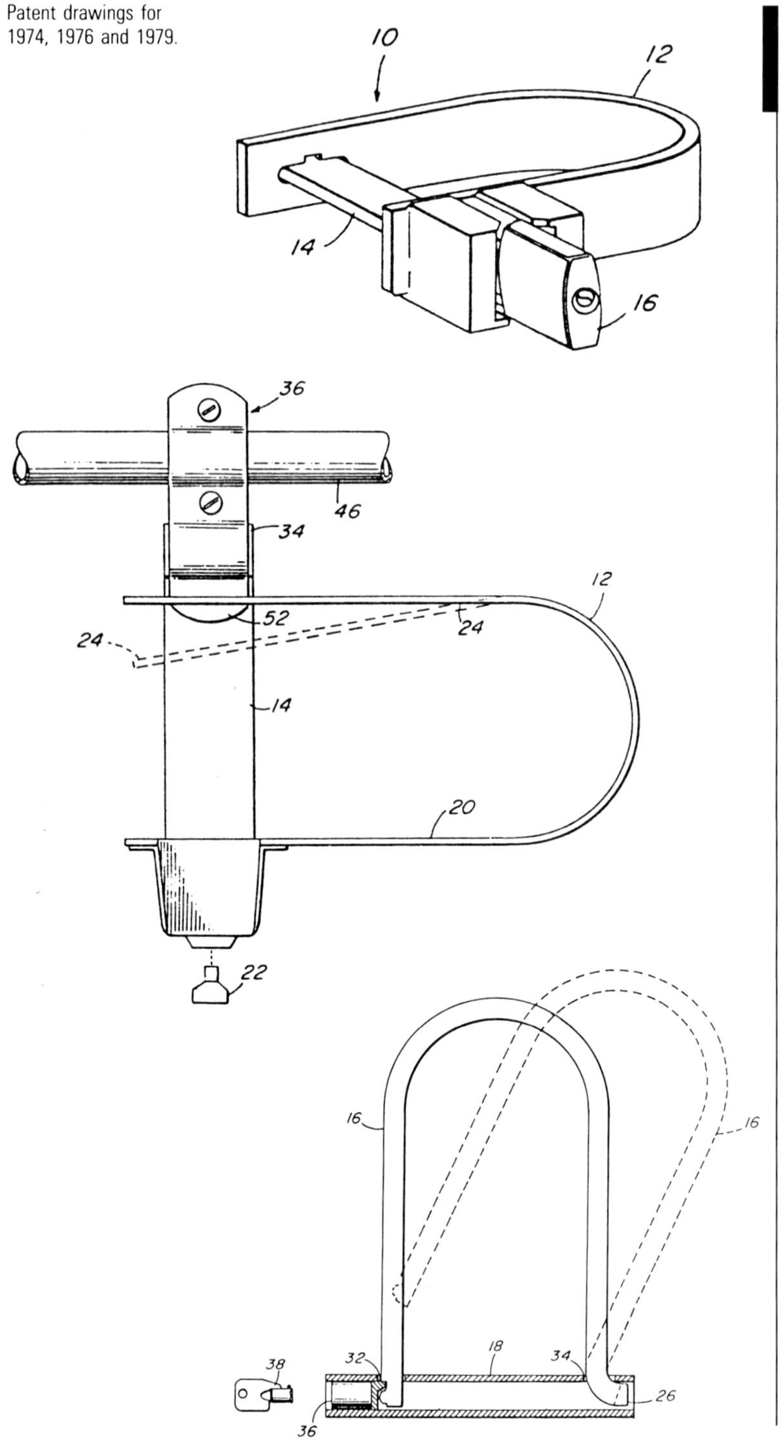

Patent drawings for 1974, 1976 and 1979.

The Design Problem: To develop a lightweight and functional lock that would withstand prying, smashing, sawing or cutting.

Lock design began at KBL in 1971. The early designs lacked practical and aesthetic qualities. "In this respect, our experience helped to define our design process. The original Kryptonite, of which only 50 were produced, looks like an Iron Age monstrosity (see model) when compared to the present design. The evolution of the design from 1971 to 1977 was a result of help from college professors, lock users, retailers and certainly our own changing and need to survive."

The moment of enlightenment in the design came with the use of steel tubing for the cross bar. This is light in weight and can be "married" to the rods with the patented "bent foot."

The lock uses state of the art ACE II tubular lock mechanism and a patented disappearing hinge, which, when locked, cannot be pulled away from the cross bar.

The bike lock is completely vinyl-coated and can be installed and removed in seconds.

This product is in the Design Collection of the Museum of Modern Art.

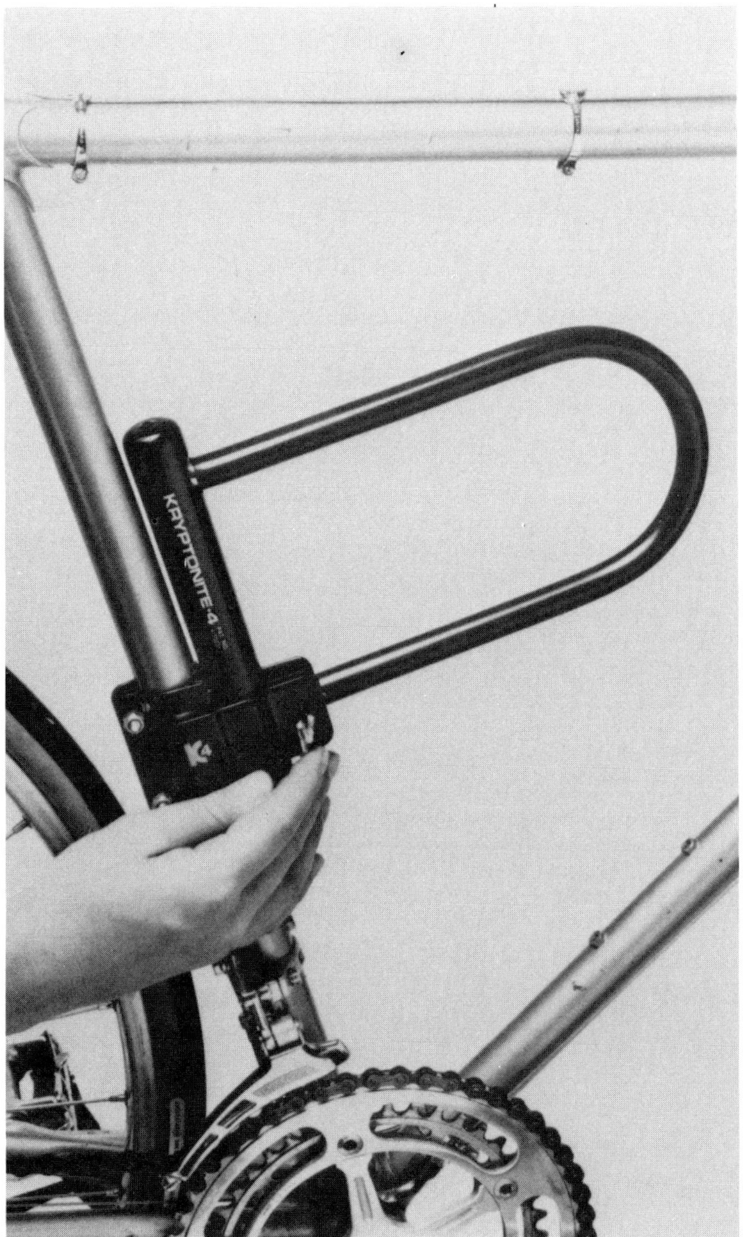

How the lock attaches to the bike frame.

Manufacturer KBL Corp.
President Michael S. Zane
Vice president Peter L. Zane
Plan manager Ernest Zane

From what looked like a Dark Ages torture devise, the product evolved from 1971 to 1977 to be selected for the design collection at the Museum of Moden Art.

Fisher Montare MountainBike 1983

The Design Problem: To provide the rider with a bike to travel back country areas, where surface consistency and elevation vary considerably.

Off-road biking began with a group of renegade cyclists in Marin County, California, who discovered the joys of "klunking" down the hillsides on old Schwinn Excelsior X bikes and exploring the fire-roads and trails in the woods.

In 1979, Gary Fisher, Tom Ritchey and Joe Breeze, all cyclists, came together to build a new, rugged frame bike specially designed for the rigors of trail riding. It was built with lighter tubing—oversize double-butted chrome-molybdenum steel tube. The new product weighed 28 pounds. Innovative features on the 18-speed Montare model include BullMoose handlebars, Hite Rite seat adjustment and flat tires.

Until 1982 almost all "klunkers" (mountain bikes) were made by hand. Now most U.S. bike companies are tooling up and Japan, England and Italy are also beginning manufacture.

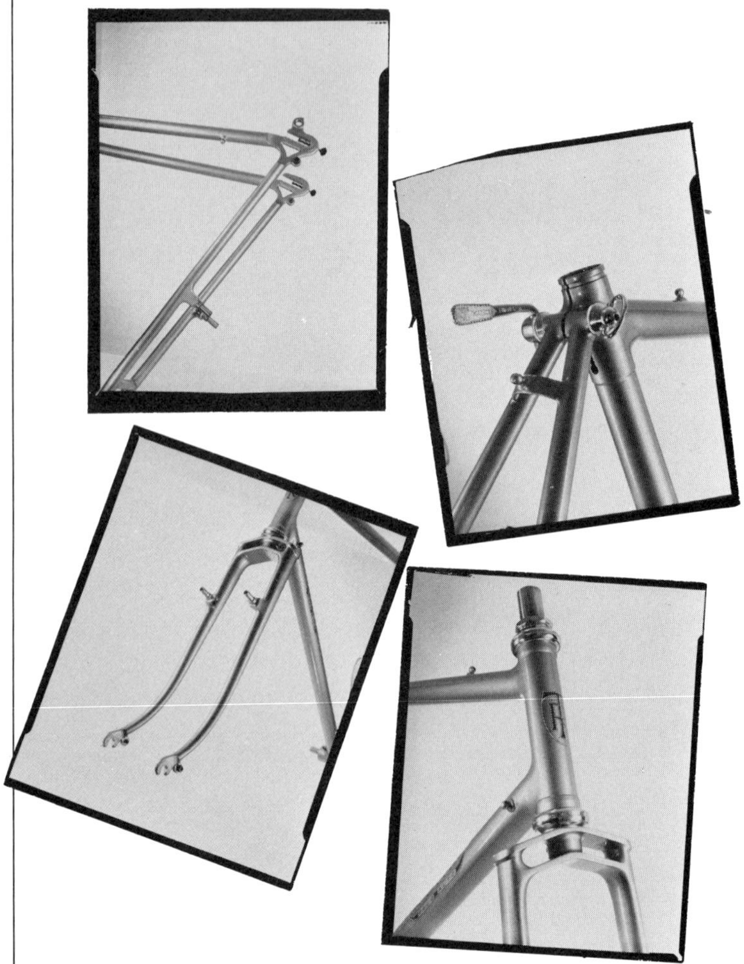

Frame is designed to withstand the rigors of trail riding.

For serious off-road biking, the 28-pound Montare has BullMoose handle bars, a Hite Rite seat adjustment, and fat tires.

Manufacturer Fisher Mountain Bikes
President Gary Fisher
Designers Tom Ritchey, Craig Mitchell, Joe Breeze

★ **Eclipse Professional** 1983

The Design Problem: To design perfectly sized bicycle packs for the serious tourist.

A lightweight (49 ounce) pack, made of cordura nylon and nylon mesh, was designed to fit almost any touring situation. It contains a large main compartment, a 72 cubic inch pocket on the outside of the coverflap, an inside map pocket and adjustable, drawstring closures, which allow for another three inches of expansion.

Manufacturer Eclipse, Inc.
President and product designer Shaun Jackson

This is a perfectly sized lightweight pack that fits almost any bicycle touring situation.

Corningware® Little Dishes 1977

Pyroceram glass ceramic was first used in radomes, the nose cone for guided missiles, and it was later brought into the consumer arena by Corning for cookware.

The Design Problem: To design new and versatile cookware forms in Pyroceram that would respond to changing lifestyles.

The story of Pyroceram glass ceramics begins in the early 1950s in Corning's research laboratory with the discovery that chemicals and controlled heat treatment applied to glass produced a tough ceramic-like material. It was unique as a ceramic since it could be melted and formed into any conventional glass shape. Moreover, it was stronger and harder than glass, more heat resistant and able to withstand the thermal shock of temperature extremes. The fact that Pyroceram is transparent to microwaves has become increasingly important with the advent of microwave cooking.

The new material was first used in radomes—the nose cone for guided missiles. A collaboration between marketing and science brought it into the consumer arena and led to the development of the Corningware line of glass ceramic cookware.

The newly discovered material was adaptable to the oven and rangetop. Twenty-five years ago, the idea of freezer-to-oven-to-tableware was revolutionary. Fire and ice demonstrations were necessary to convince a skeptical public that a cookware material could withstand such temperature changes.

Throughout its history, Corning has taken a customer oriented approach to business. Using consumer focus groups and market research, the company continually monitors changing preferences and cooking trends. New products are introduced and old products discontinued as a result of these studies.

Little Dishes meet the basic cooking and eating needs created by changing family and household structures and eating habits. They were designed for cooking and serving individual portions for those eating alone.

Corningware® French White™ Cookware 1978

French White was developed to appeal to new customers who were seeking "upscale" or gourmet cookware and to complement a variety of kitchen and dining environments. The simple forms and plain white body meet the need for classic, universal forms in a time of rapidly changing lifestyles.

Consultant designers Davin Stowell Associates
Designers Davin Stowell, Tucker Viemeister
Manufacturer Corning Glass Works, Consumer Products Div.
Project designer Richard W. Greger
Project engineer Earl J. Whitney
Project engineers Charles Benardo, Creighton Lawhead, Dennis Naylor

The French White line (top) is gourmet cookware. The Grab-A-Snack set (left) and Grab-It bowl (right) are casual and adapt to oven and rangetop.

Lauffer Design 10™ LEXAN® Flatware 1982

The Design Problem: To design flatware of a larger scale than any that had been previously produced. In designing Magnum, Wallance wanted to develop a new profile with a distinct articulation of three sections. His objective was to achieve a fusion of visual beauty and tactile quality.

Wallance's first flatware design for Lauffer, *Design One,* appeared in 1954. As in subsequent designs, his goal was to seek forms that could not only be mass-produced, but would also express the stamping and forging processes involved and the enduring quality of the metal itself.

Wallance considers himself an industrially oriented craftsman, aware of the market, yet needing to identify with the product he designs. He states that his designs are 20% conception and 80% refinement. Attention to the most minute subleties is essential in flatware design and "makes the difference between ordinary design and one that sings."

Don Wallance's early models illustrate form studies.

The designer moved away from designs characteristic of stainless flatware to the simple, sturdy form appropriate to plastic.

Designer Don Wallance
Manufacturer Lauffler (A Towle Company)

★ **Screwpull Corkscrew** 1979

Engineering drawing.

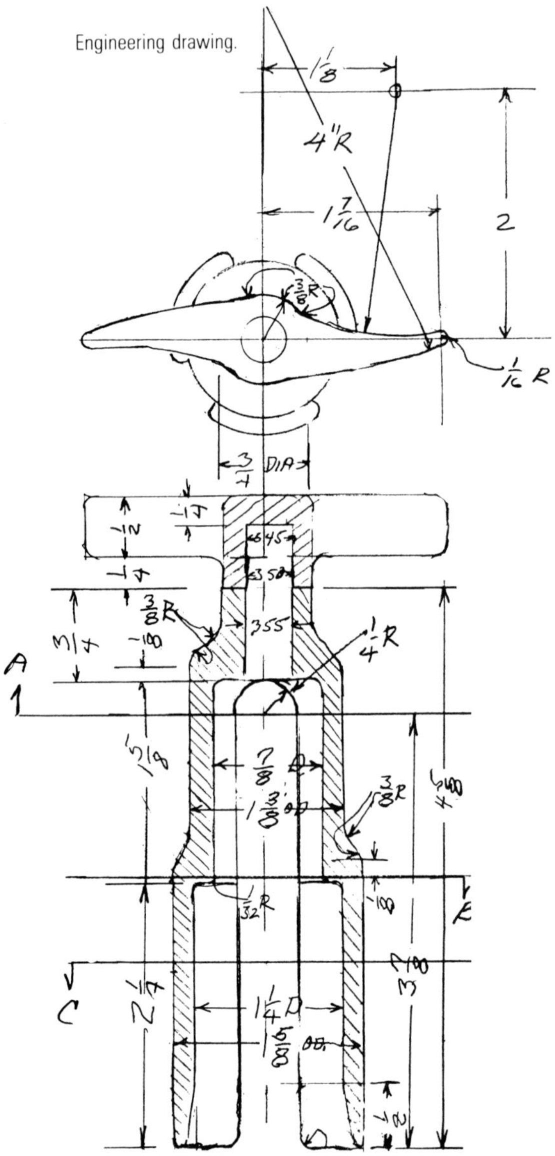

The Design Problem: To develop a screw operated corkpuller for opening both tough, hard corks and fragile, crumbling ones from vintage bottles.

Herbert Allen, inventor and designer, has spent 50 years in the oil business "inventing" drilling devices for oil rigs. The difference between that and bottle corks seemed to be largely one of scale. In 1950, Allen started a wine collection and began to consider the corkscrew. It was quite a challenge in a business as old as corkpulling to come up with something completely new.

The design process took two years. "It had been my procedure in developing a great many inventions to perform experiments, including working models of an idea before doing research on the proposal—early study of the area might limit the scope of originality."

Therefore, after constructing early models, he visited the Library of Congress, N.Y. Public Library and Westminster Library in London; he also discovered the International Correspondents of Corkscrew Collectors. After this research and examination of patent files, he concluded that his design was superior to the published art, original and patentable models.

The pivotal moment occured with the use of an anti-friction (teflon) coating on the screw. Allen was familiar with the remarkable properties of teflon, having used it in the development of oil field equipment. The design consists of a simple plastic guide frame that settles down on the wine bottle, plus a thin, helical screw of high tensile, teflon-coated stainless steel which pierces the cork right through the center.

The Screwpull was developed from concept through manufacture to world-wide commercial distribution by the designer. It is part of the Design Collection of the Museum of Modern Art.

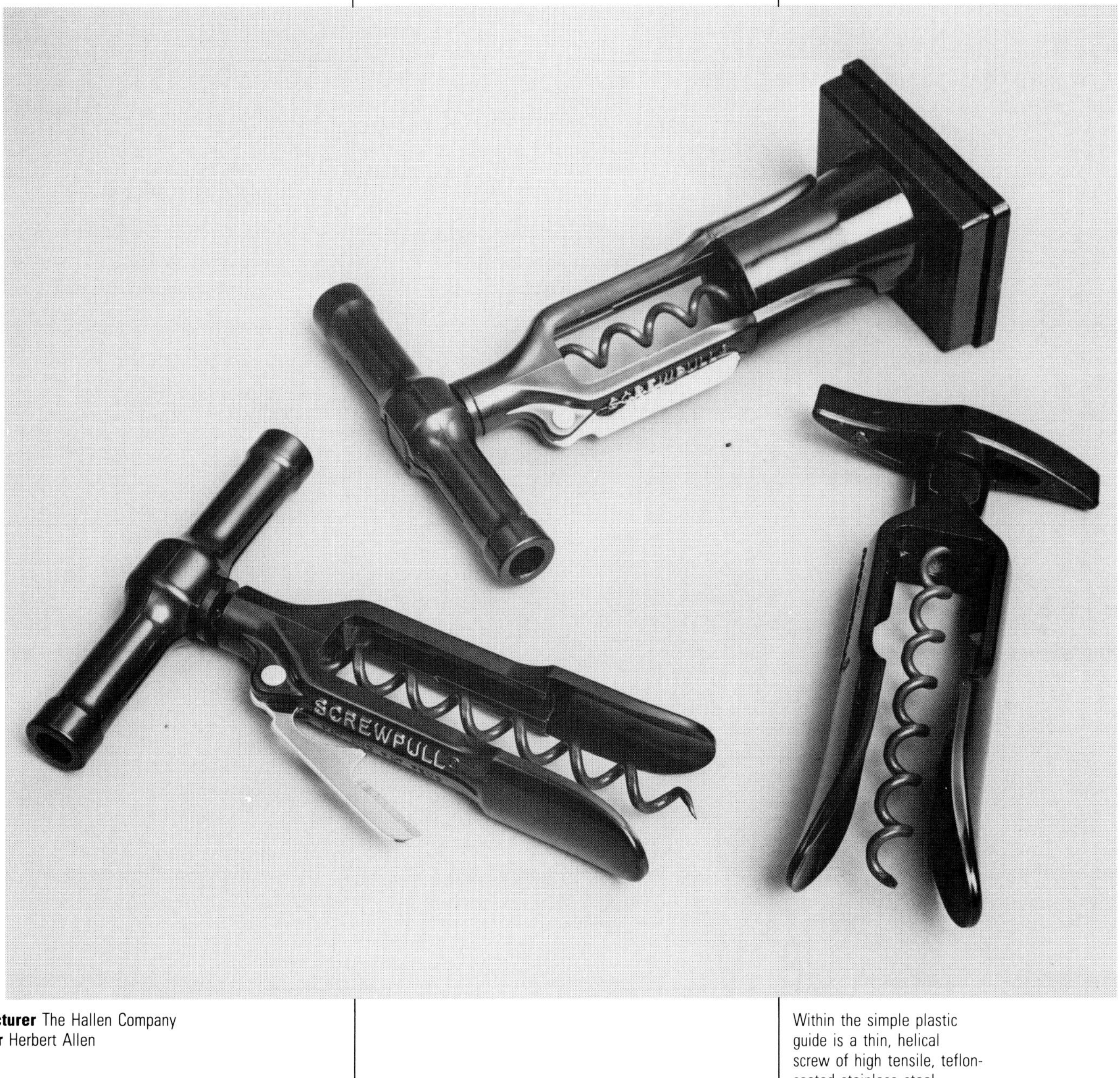

Manufacturer The Hallen Company
Designer Herbert Allen

Within the simple plastic guide is a thin, helical screw of high tensile, teflon-coated stainless steel.

AMERICAN

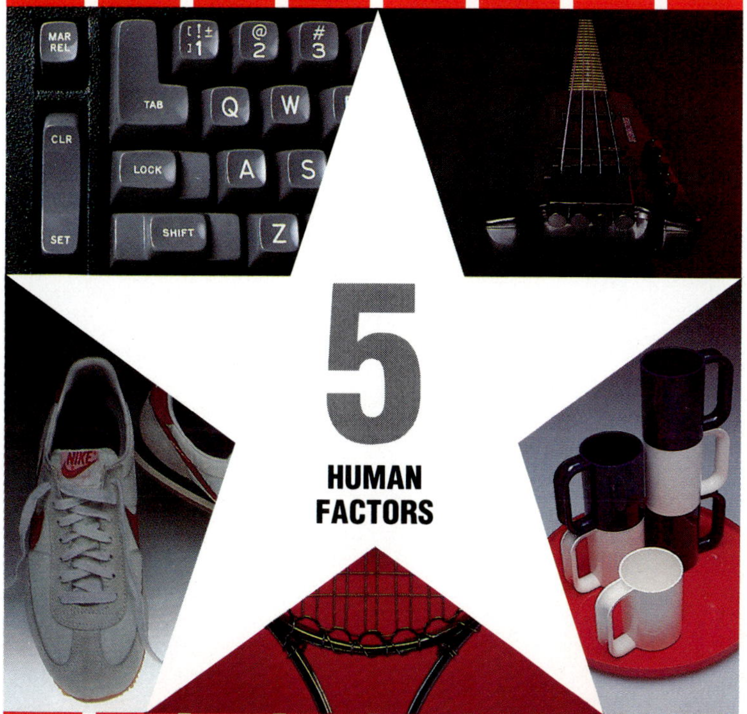

DESIGN CLASSICS

Lamb Wedge Lock Handle for Cutco Cutlery 1952

Studies that led to the development of the Wedge Lock Handle.

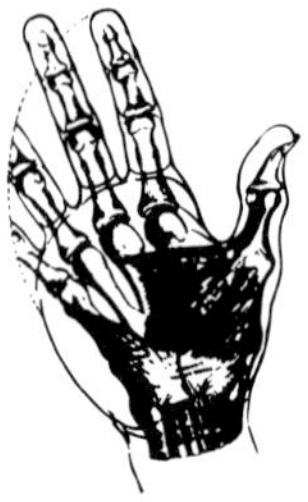

The human thumb.
Studies in strength, mobility and grip resulted in the discovery that the full power of the thumb had never been utilized in traditional handle design.

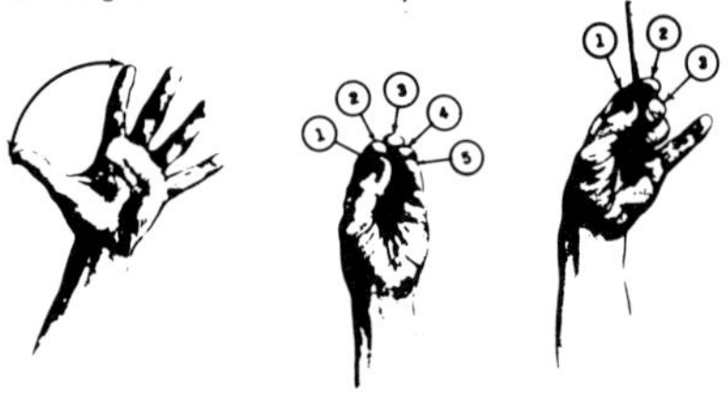

Comparative strength of the fingers.
The thumb and forefinger of the hand have two-thirds of its strength, yet handles were designed so that the three weakest fingers assumed the work load.

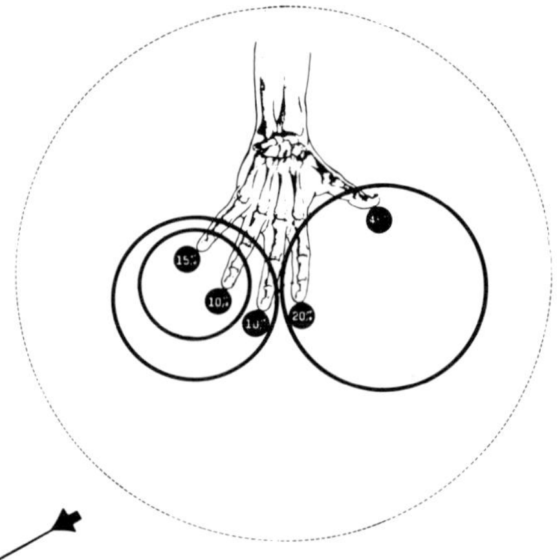

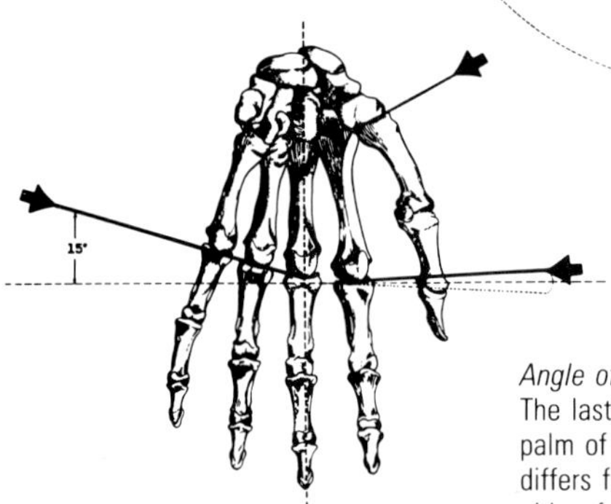

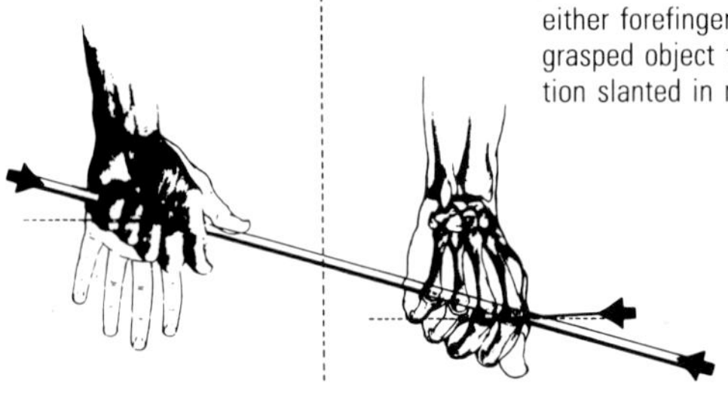

Angle of the hand.
The last three fingers are joined to the palm of the hand at an angle which differs from the angle of attachment of either forefinger or thumb. Thus a grasped object falls into a fixed position slanted in relation to the arm.

The Design Problem: To design tool handles that used the strengths and flexibilities of the human hand.

From the vantage point of 86 years, Tomas Lamb now believes that his entire life, since childhood, has been directed to the study of the human hand. It was not until World War II, however, that Lamb abandoned a successful career as a textile designer to devote his full attention to this interest.

Six years of extensive research and experiments led to a better understanding of the anatomy and function of the hand and resulted in a new concept of efficiency in handles. The Wedge Lock Handle derived its form from a system of scientifically designed contours, which utilized the full power of the thumb, forced every muscle and tendon to work at their maximum ability without straining beyond their natural limitations, fully equalized the work load on different fingers of the hand, and applied the load at the proper angle for each finger. The grooves wedged the fingers apart over the entire handle and locked the thumb and fingers in place for a strong firm grip. The curves were designed so that the handle can be operated at any angle.

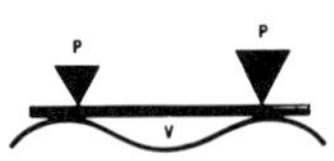

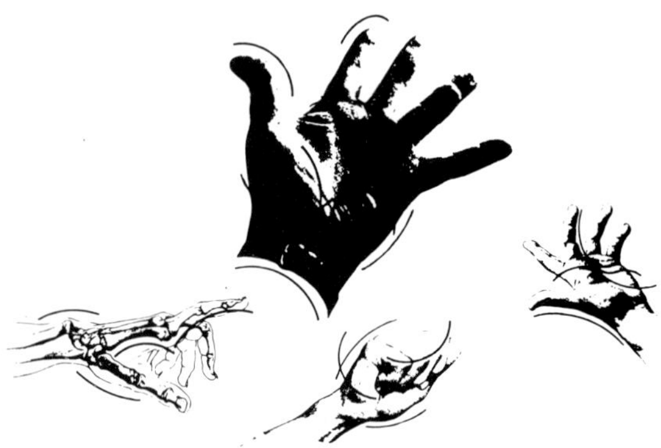

Curves of the hand.
Flat sided objects meet these curves only at high spots, allowing vibration, which causes lack of gripping strength.

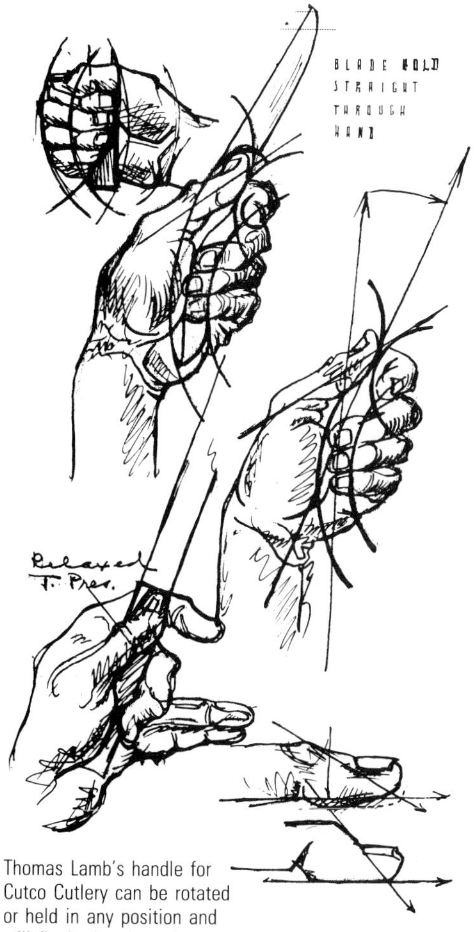

Thomas Lamb's handle for Cutco Cutlery can be rotated or held in any position and still fit the hand comfortably.

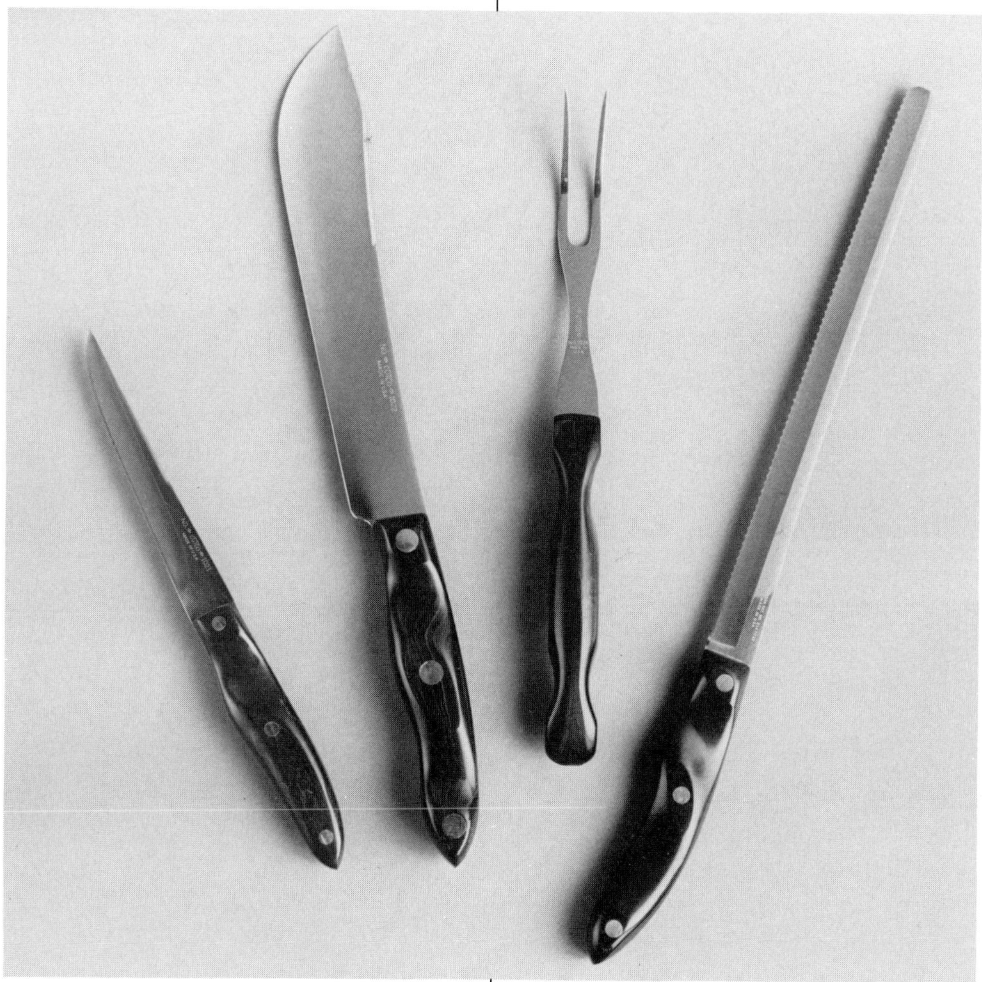

Designer Thomas Lamb
Manufacturer Alcas Cutlery Corp.

Lamb Universal Wedge Lock Handle for Cutco Cutlery 1971

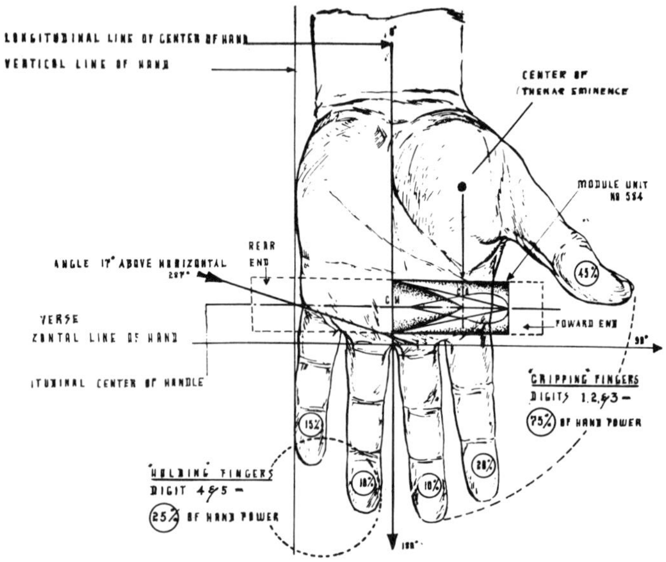

The first handle invention used gravity and solved the problem of handling weight in downward and upward pressure. The Universal Wedge Lock Handle, which was incorporated into Cutco Cutlery in the early Seventies, works in a sphere. The result of more than 10 years of study, the latest design involves the handling of gravity in any direction. The handle can be rotated or held in any position and still fit the hand comfortably.

Archaeologists have traced the handle back at least a million and a quarter years. Lamb contends that "human civilization began when the human animal happened to seize his first handle." It wasn't until Lamb's scientific study of the hand led to change in the basic form of the handle that a major improvement in facility and comfort in use was achieved.

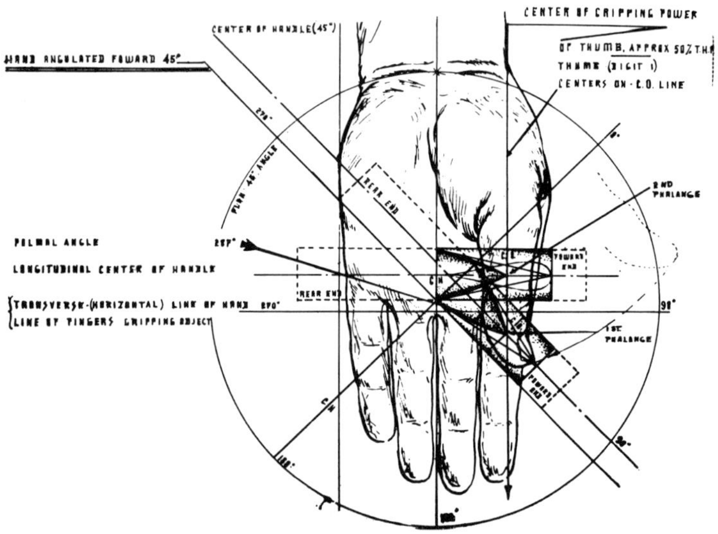

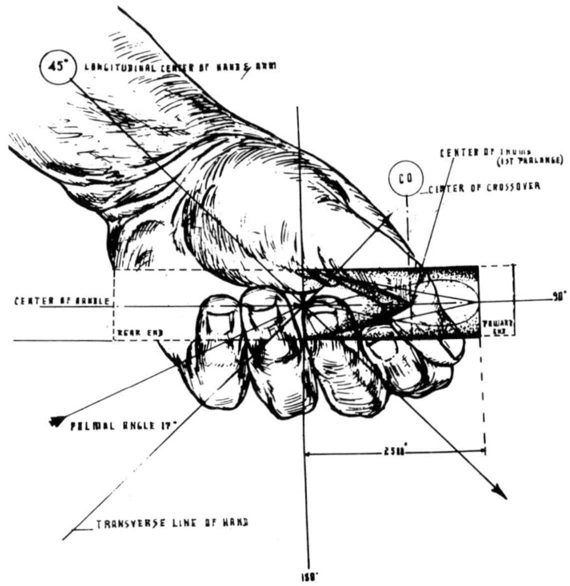

Studies that led to the design of the wedge lock handle.

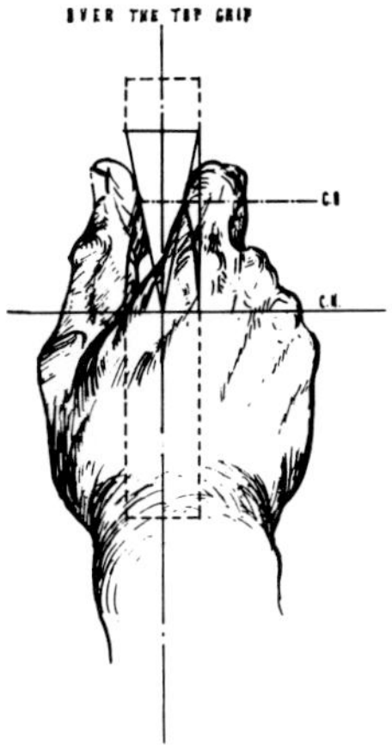

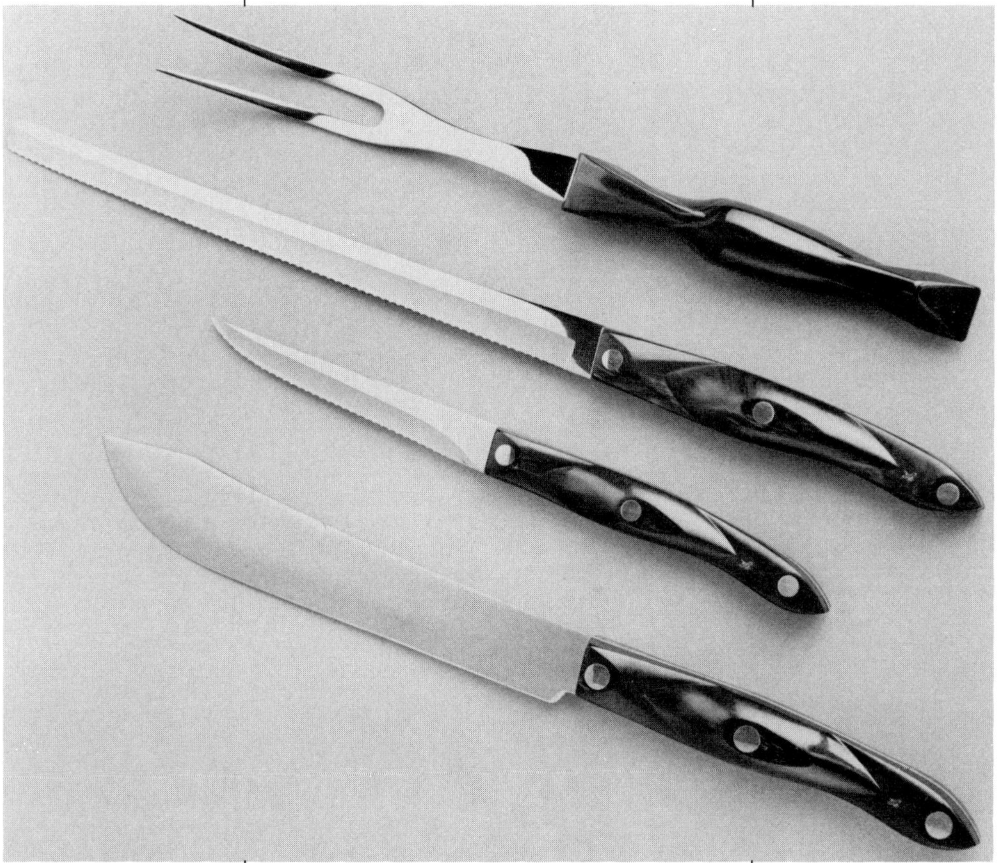

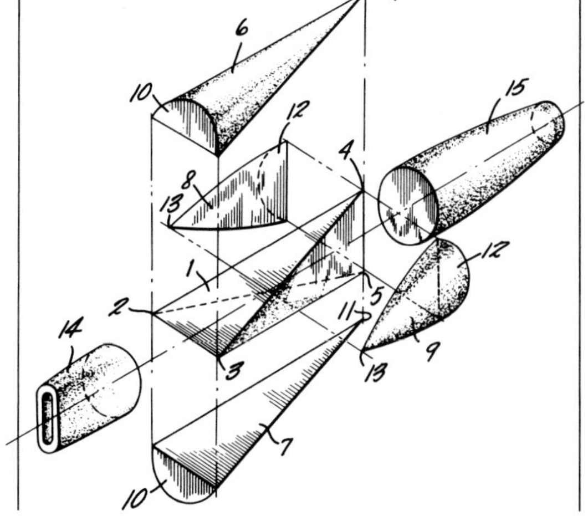

Designer Thomas Lamb
Manufacturer Alcas Cutlery Corp.

Chicago Cutlery BioCurve Knives 1983

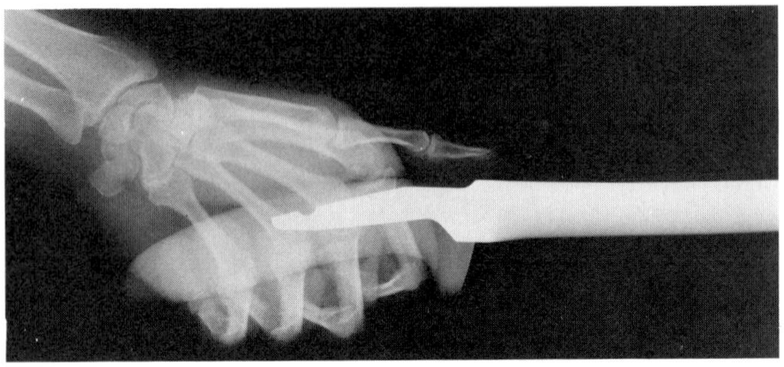

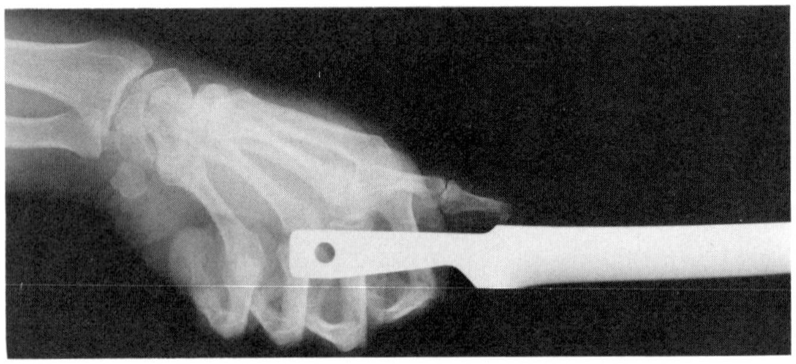

X-ray photographs of standard (top) and BioCurve form (bottom) illustrate more comfortable position for hand and wrist.

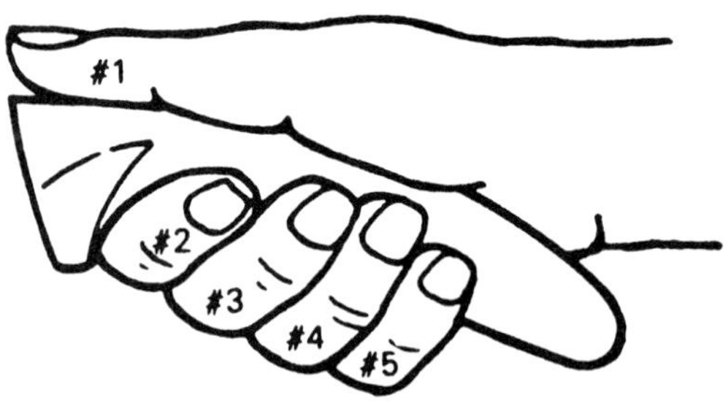

The Design Problem: To improve the ergonomic function of handles.

This innovative handle, called "Bennett's Bend," is the work of John Bennett, who discovered in 1972 that tools, toys and utensils are easier to use if they are built with bent handles.

According to Bennett, the ideal position for the hand when holding a tool is the neutral position. The angles formed by the naturally clenched hand are the basis for the new handle curve and shape he has designed.

Studies have shown that when you use a straight handle, the wrist must be bent. This bent wrist, or "ulnar deviation," causes stress and tension in the wrist and forearm muscles and reduces the range of wrist motion. Tests also show that blood pressure is increased due to positioning of the tool.

By bending the handle 19°, the bent wrist is eliminated. The improved, bent handle allows the user to grip the tool with the two strongest fingers.

The handle top is forced against the ball of the thumb, locking the handle in the hand. When the BioCurve(TM) knife is gripped, the middle finger fits the bottom curve of the handle. The remaining fingers fall naturally and comfortably into place, and the wrist is straight.

The American Chef knives featuring the patented Bennett's BioCurve handle are being used in meat processing plants where wrist stress has been a chronic problem.

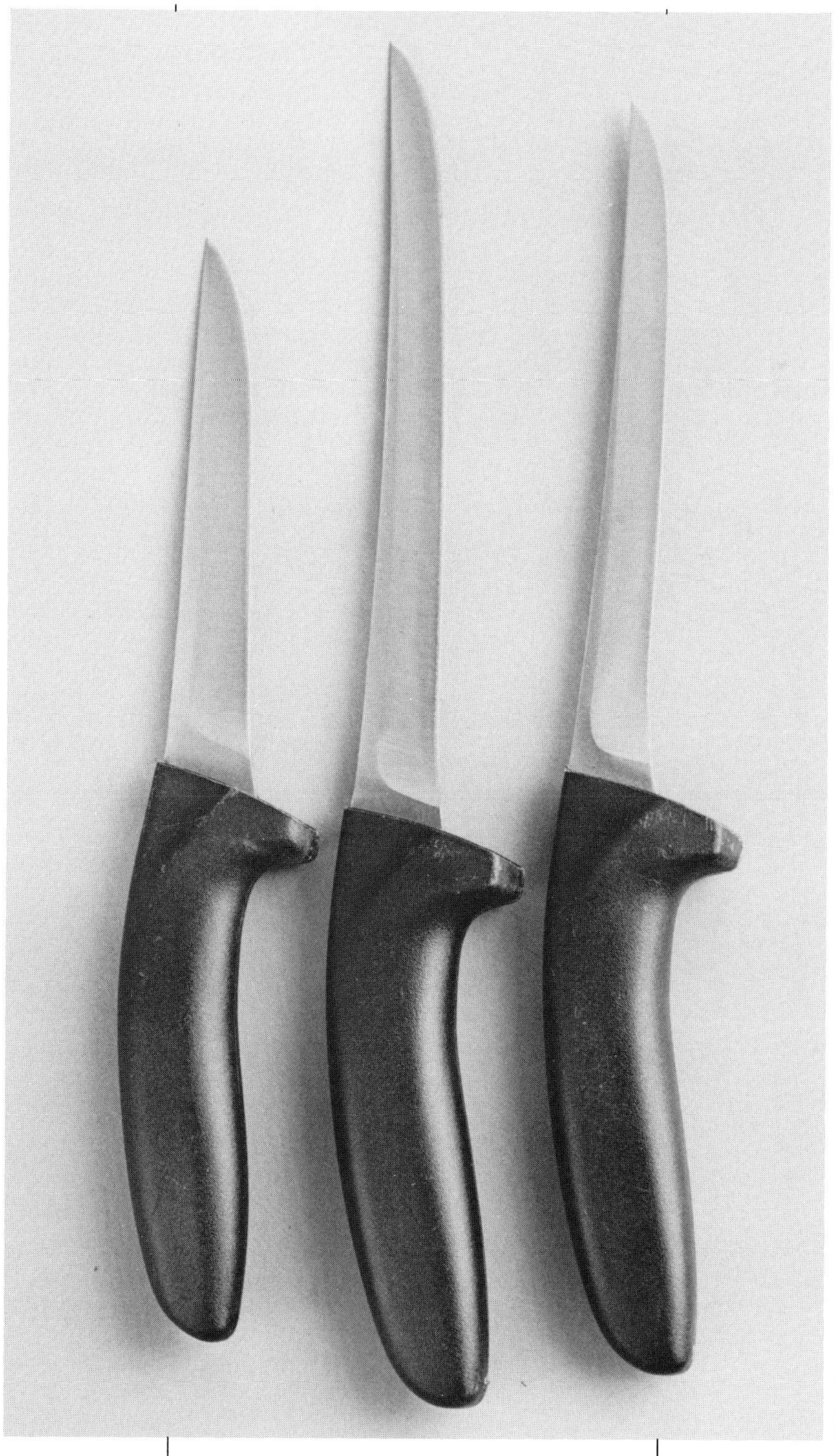

The Bennett Bend handle design applied to a set of knives.

Manufacturer Chicago Cutlery
Designer John Bennett

Stanley Saw Handle 1981

Drawings of early concepts.

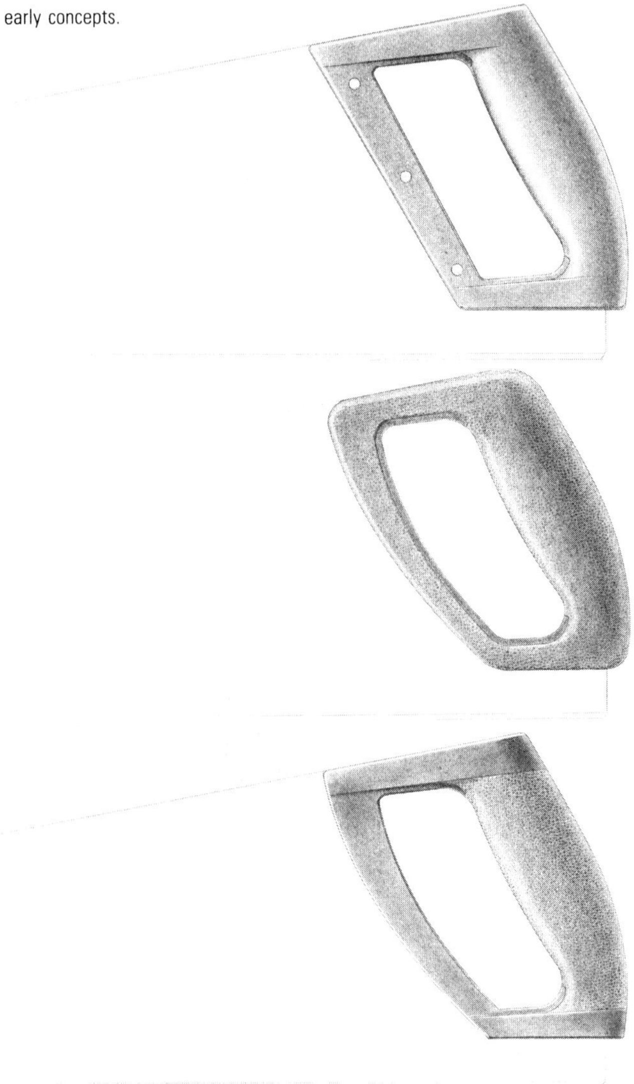

The Design Problem: To design a molded plastic saw handle that would comfortably fit the human hand, have an appearance appropriate to a cutting tool and, in this case, comply with the requirements of the welding process in assembly.

The typical wood handle has an inward negative curve in the finger grip area. Since the human hand has the longest fingers in the center, the maximum girth should be in the center of the handle.

After preliminary concept studies, the most important stage in handle design begins with the making of a sculptured plasticene model. Working in this way, the designer is able to grasp and evaluate the shape while modeling, making subtle refinements to increase handling comfort.

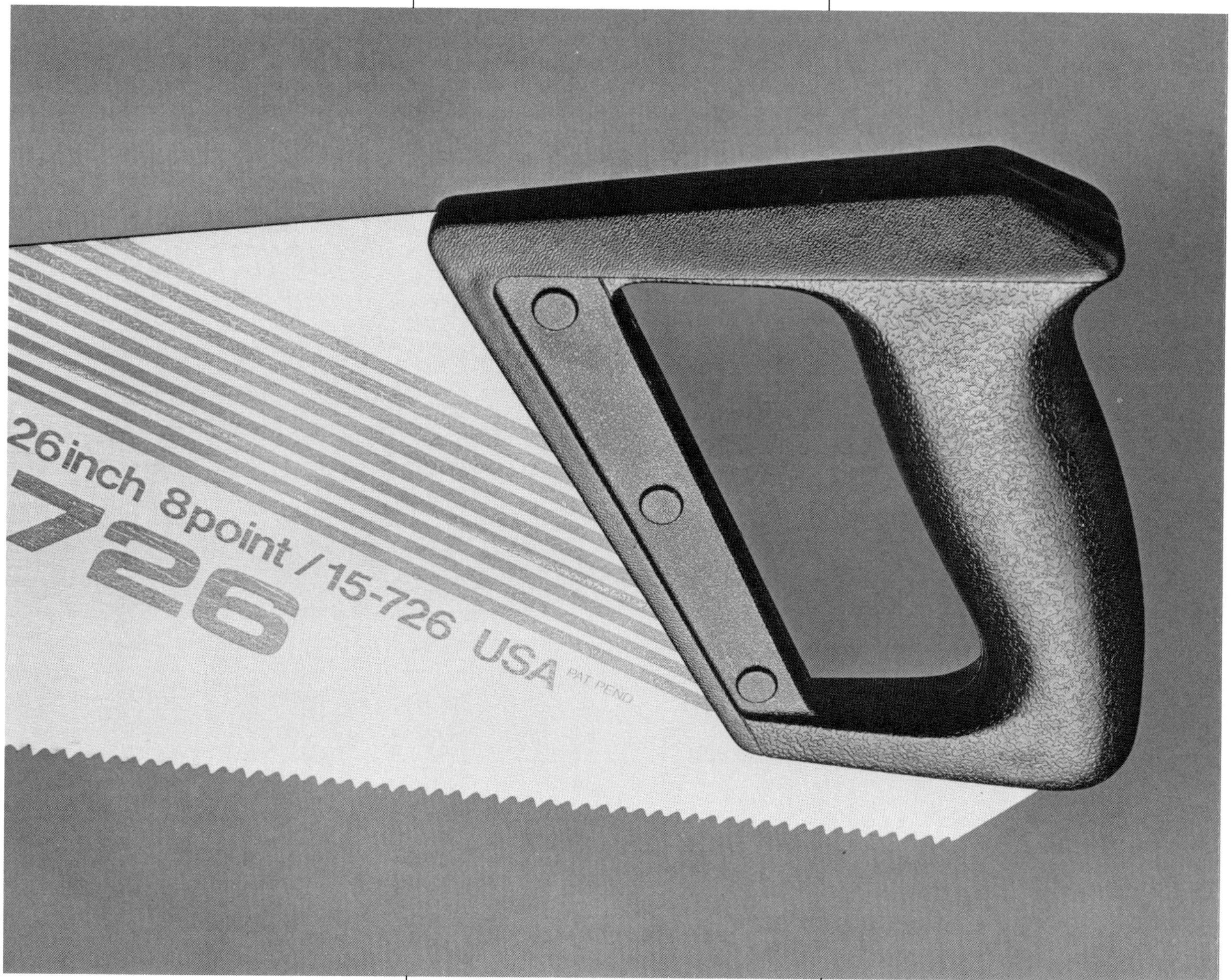

Consultant designer Laird Covey
Manufacturer Stanley Tools, Div. of Stanley Works
Manager of new products research Robert F. West
Manager of product development Walter Hutchins

The physical properties of the hand were carefully considered in designing a new handle.

The Gripper™ Tennis Racquet 1982

The Design Problem: To maximize human potential and minimize exposure to trauma.

The major purpose of racquets is to act as a transmitter of force from the human body to the ball. The Gripper™ is ergonomically designed to conform to the natural contours of the hand. The need to grip the handle tightly is eliminated, adding as much as 50% increase in strength. This new handle also eliminates the need to change grips from forehand to backhand or when serving.

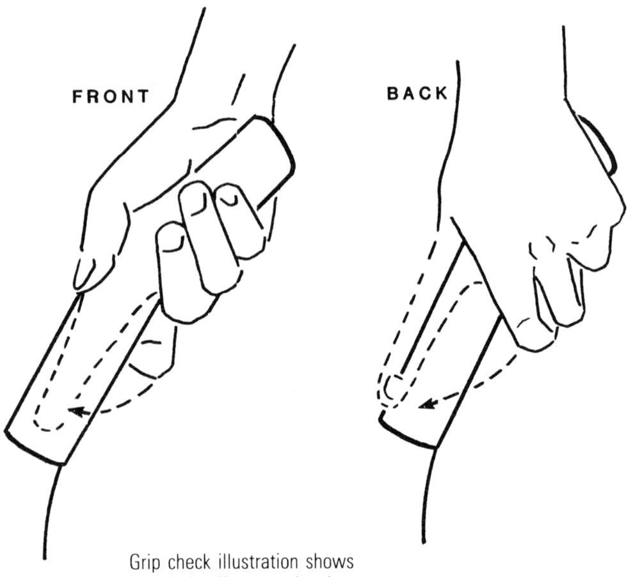

Grip check illustration shows how index finger or thumb are held parallel to the handle for forehand or backhand stroke.

Based on ergonomic principles, the racquet is designed to maximize human potential and minimize exposure to trauma.

Manufacturer BioSports, Inc.
Designer John Bennett

NIKE Odyssey Running Shoe
1982

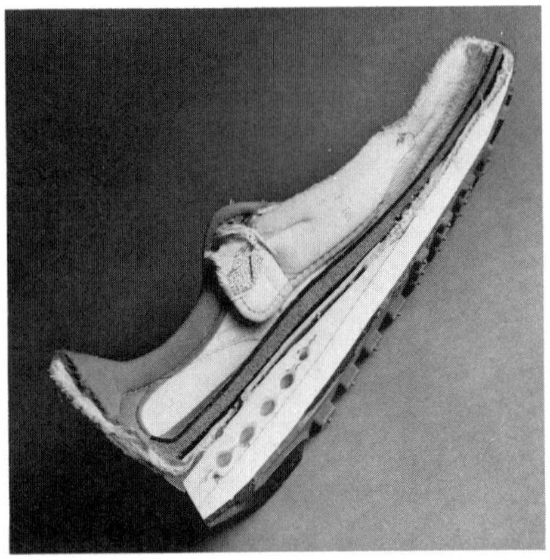

Cut-away shows sole cushion and upper section.

Details of the tread layout.

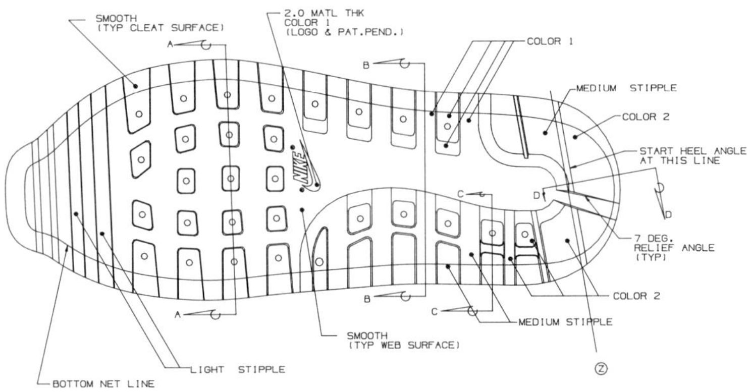

The Design Problem: To create a training shoe that combines exceptional stability with superior cushioning.

Well-cushioned running shoes are often unstable and allow excessive pronation of the foot. Those that provide stability and limit pronation tend to be firm or poorly cushioned.

Research into last design proved that the wide medial arch, originally conceived for leisure and walking shoes, would be the solution to the stability problem of the running shoe. By using this last, which is completely straight from heel to toe, the designers were free to use any cushioning system they desired to decrease leg shock. A "center-of-pressure" outsole design and the patented external heel counter support aid in the cushioning and stability of the shoe. The NIKE Air-Wedge cushioning system is used in the heel.

Destiny is a new running shoe for children. It is the first youth shoe to systematically adjust the height of the mid-sole and wedge according to shoe size, so that each size receives the amount of cushion it deserves.

Manufacturer NIKE, Inc.
Design NIKE Sports Research Laboratory
Director Dr. E.C. Fredrick
Biomechanics Dr. Tom Clarke
Consultant John Hagy
Advanced concepts director Bill Peterson
Advanced concepts Jim Crawford, Tom Derderian
Project managers running Bill Mintiens, Dan Norton
Line development manager Mark Parker
Air development managers David Stearns, Pete Warzinski
Director corporate technology Jeff Johnson
Director product development David Knottkamp
Air technology consultant Frank Rudy

NIKE Skate Nordic Ski Racing Shoe 1983

The Design Problem: To design high performance ski racing boots that maximize turning control of the skis and the skiers' stability.

The outsole design of the Norwegian inventor, Kjell Aarflot, was the key element in the new design of the Nordic boot, which is the first to incorporate bindings, boots and skis as a simple, integrated system. Each component form relates directly to its function in the boot. The new design is an improvement over the traditional Nordic boot, since the full sole of the boot mates with ski all the way from toe-clip to the heel. It eliminates the need for control plates attached to the skis. The sole is textured for improved traction. (The boot is mounted here to show how it relates to the ski, simply and directly).

Human factors and function were the major considerations; the boot needed to give ideal fit and allow the athlete to use his energy optimally. The materials were critical, because they needed to endure the excessive stresses of competiton and the unusual demands of the cold and moisture experienced by a skier.

The NIKE approach to product development combines the talents of outside consultants and in-house teams of specialists in various fields of shoe design.

Design consultant Kjell Aarflot
Manufacturer NIKE, Inc.
Project manager emerging sports Kevin Gosney
Line development manager Trip Allen
Industrial design Ken Greer
Advanced concepts director Bill Peterson
Advanced concepts Tom Derderian
Director corporate technology Jeff Johnson
Director product development David Kottkamp

The boot mounted to show how it relates to the ski, simply and directly.

Bell Tourlite Bicycle Helmets
1982

The Design Problem: To design a bicycle helmet that offers protection, yet is light-weight and cool.

Statistics prove that bicyclists can avoid serious head injury in an accident by wearing a helmet. Bell helmets are engineered to provide protection by reducing head acceleration during impact and by absorbing as much of the impact energy as possible. In testing, they meet or exceed national and international performance standards.

The outer shell of the Tourlite helmet is of a molded high impact PVC plastic, lined with a one inch thick expanded polystyrene to protect against impact. Additional safety features include an adjustable tinted visor to reduce glare and reflective striping to increase night time visibility. Aerodynamically designed air inlets channel a continuous stream of fresh air through the lining for cooling and ventilation.

Interior of helmet reduces head acceleration during impact and absorbs as much energy as possible.

Manufacturer Bell Helmet Company
Department head industrial design James Sundahl
Industrial designer Lester Broersma

The outer shell has tinted visor and air inlets to channel a stream of fresh air through the lining for cooling and ventilation.

★ Motorola Optrx Display Data Radio Pager 1982

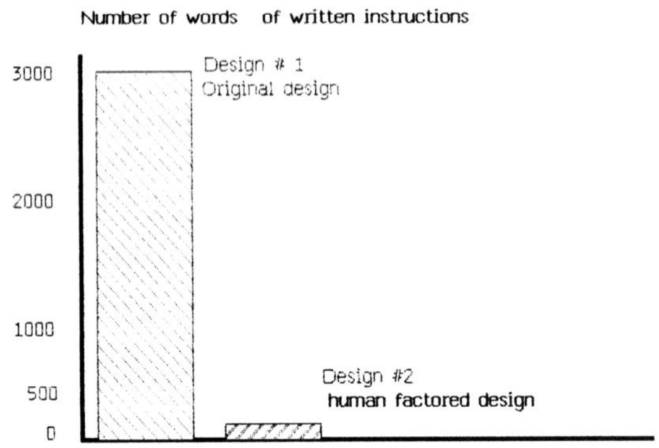

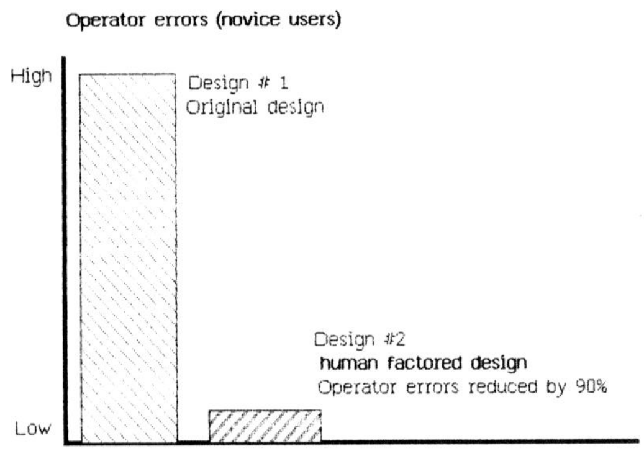

With new design, errors committed by novice users was cut by 90%.

The Design Problem: To develop a product that offers the user far more capability than previous models and is easy and self evident to use.

The display pager represents an advanced, one-way communications, personal message receiver that fits into a pocket or in a belt-clip case.

Research, employing the User Merit Index (developed by Mauro Associates), made it possible to determine the operating difficulties of various engineering prototypes. Computer simulations of controls and display were used to arrive at an overall rating of several designs. The UMI also provided detailed directions for the design of a new configuration that would meet the needs of the user group.

There was a reduction in the number of written instructions from approximately 3,500 words to 75. The percentage of errors committed by novice users was reduced an estimated 90%.

Motorola's industrial design and engineering groups played a key role in the development of the product; Mauro's office was responsible for human factors and reducing operational complexity.

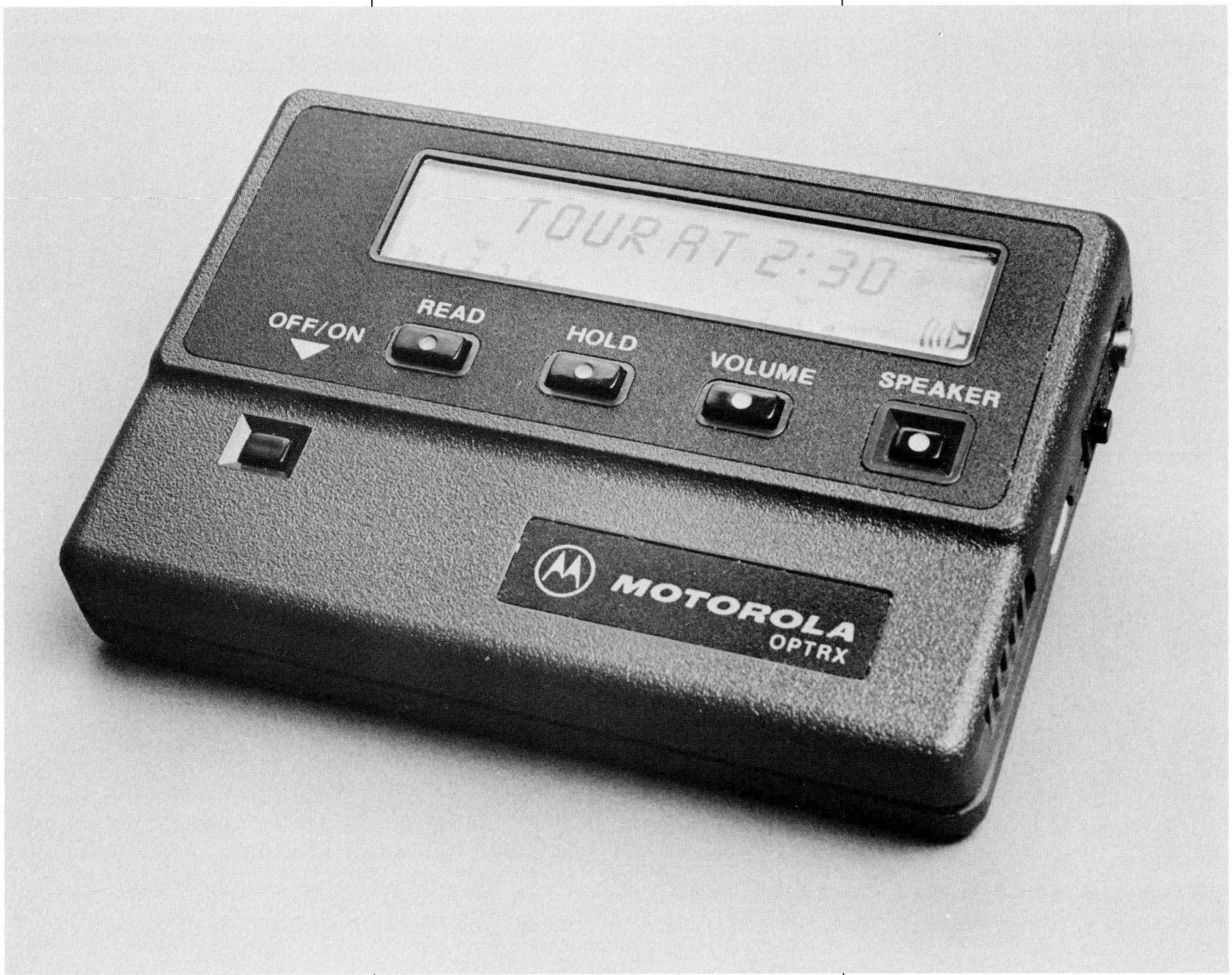

The designer's role was to be responsible for human factors and reduce operational complexity.

Consultant designers C.L. Mauro Associates, Inc.
Manager Charles L. Mauro
Program director Dr. Hugh M. Bowen
Manufacturer Motorola Portable Paging Products Div., Motorola Company
Manager Don Drumin
Designer Richard Toth
Manager product engineering Robert Swendeman

Cuisinart Food Processor 1981

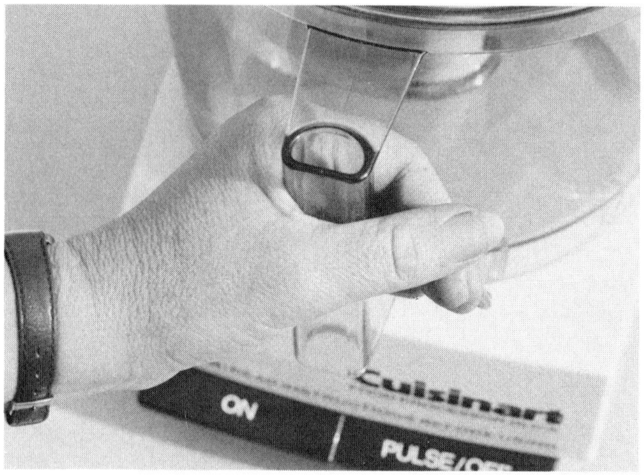

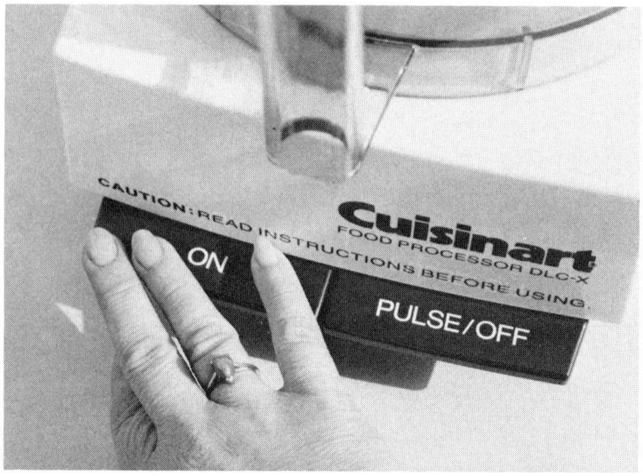

The Design Problem: To develop a new, larger capacity food processor with improved user features, human factors, ergonomics and safety considerations.

Using the product over time helped shape attitudes and philosophy. The client had stated a "nothing but perfection" attitude and emphasized quality of form and detailing.

In redesigning the DLC-X, Harrison addressed the needs of the handicapped. The final design benefited this group and also provided a more efficient unit for the general consumer.

A thorough human factors study was made of the operational sequences with a review of every part, detail and form in terms of safety considerations. Research was made into the problems of those with arthritis, Parkinson's disease and visual deficiencies. Data from the Underwriter's Lab and other testing requirements played a role in the design of every element, including larger and more prominent warning labels, colors and textures.

The proportions and special shape of the work bowl handle (1) was developed from anthropometric data: the grip shape distributes the forces over a large portion of the palm. Large paddle-type controls are designed for gross hand motions for persons having manipulation problems; any part of the hand pushing upwards will activate the controls (2). Light lettering placed on the dark field permits better visibility. The shape of the plug conforms to the shape of the finger tips.

Craig Claiborne, writing about the Cuisinart food processor in *The New York Times,* said, "Its invention, in the minds of serious cooks, ranks with that of the printing press, cotton gin, steamboat, paper clips and Kleenex. . . ."

Close ups reveal the designer's care in creating and locating controls; for example, large paddle type controls are designed for major hand motions.

Design consultant Marc Harrison
Manufacturer Cuisinarts, Inc.
President Carl Sontheimer

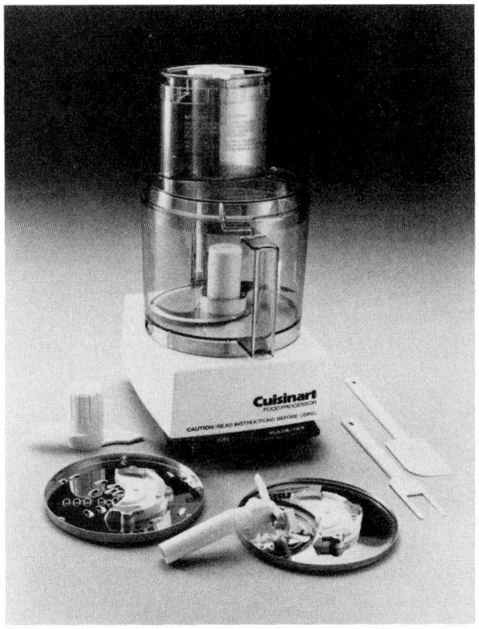

In addressing the redesign, Marc Harrison considered the needs of the handicapped, which yielded an overall better, even more attractive design.

Libbey Perfect Stackers Glassware 1979

Concept sketches illustrate the ergonomics of the design.

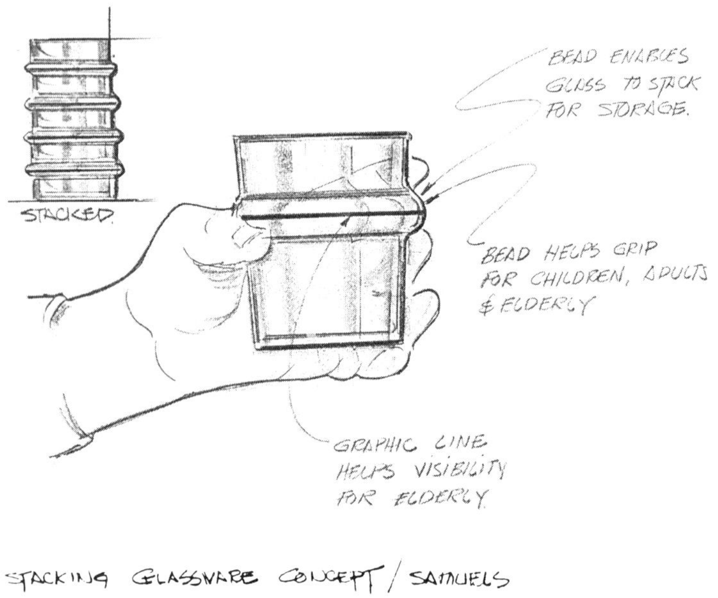

The Design Problem: To design a line of glassware appropriate to children (small hands) and the elderly (arthritic hands) that would provide easy gripping.

Allen Samuels recognized the need for this product and, when approached by his client to design a new line of glassware, he suggested a stackable, inexpensive product, but one that would address the needs of special groups—the elderly and children as well as the general consumer.

At this time, Samuels had over fifteen years experience in designing for the handicapped, working with anthropometric studies of the hands and reviewing the problems of the aging. By applying this knowledge to the design of a line of glassware, he was able to meet the needs of these special groups and, in addition, to create a pleasing and practical form.

The designer regrets that the line of glassware was never promoted for its unique human factors solutions. The product was on the market for less than two years because sales volume was under the 200,000 dozen a year required for success.

Samuels believes that he has "an obligation, as a designer, to seeing and analyzing and shaping objects in a most serious, intelligent uplighting and responsible way. You are helping to shape future generations as you communicate your ideas, discoveries and feelings through your designed objects. Good designs are those that are appropriate to a particular people, their purpose, place and time."

Manufacturer Libbey Glass
Industrial designer Allen Samuels

The design addresses the needs of special groups, especially the elderly and children.

Xerox Animal Crackers Reading Kit 1982

Learning unit in closed position.

The Design Problem: To design a set of teacher's aids that would incorporate the concepts of the National Educational Association's kindergarten curriculum in reading readiness.

The designer worked closely with educational, marketing and design departments and with outside consultants over a two-year period to establish a series of words, images and objects that would effectively meet the objectives of the curriculum.

The Gersin office provided over 500 individual graphic images, which then had to be selected, designed, coordinated; and the learning components were then studied to insure clarity.

The graphic material was printed on durable card stock with a washable plastic surface and rounded corners for safety and damage reduction.

It became evident from classroom observations that the materials should be enclosed and clearly organized in their own container. The designers developed a container that sits on the floor at a child's line of vision and opens like a book to display the entire program. Materials are located in clearly identifiable compartments so students can put everything away correctly. The molded ABS plastic container is easy to clean, safe and durable. The kit weighs less than fifty pounds and has a recessed handle so that it can be carried like a suitcase.

Design consultants Robert P. Gersin Associates, Inc.
Design director Robert P. Gersin
Product designer Daviel Murphy
Graphic designer David Curry
Manufacturer Ginn and Company, A Div. of Xerox Corp.

The unit sits on the floor (skates indicate scale) at a child's line of vision and opens like a book.

Ledu Task Light 1983

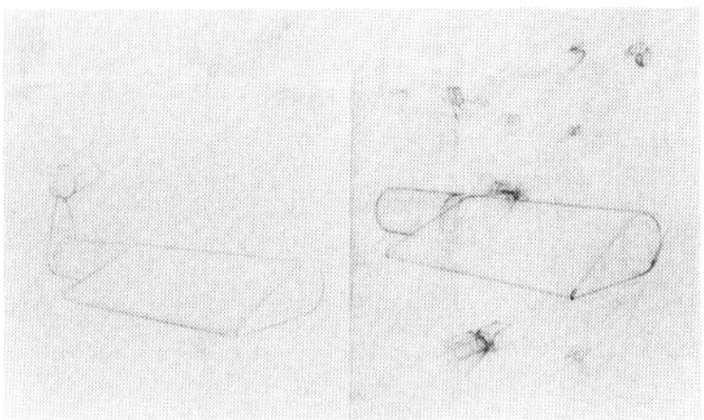

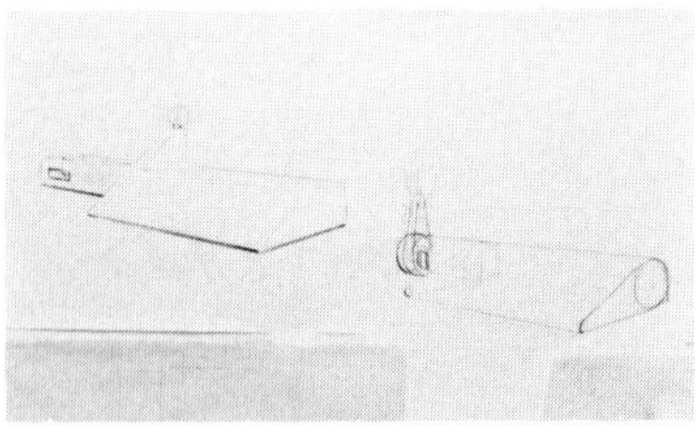

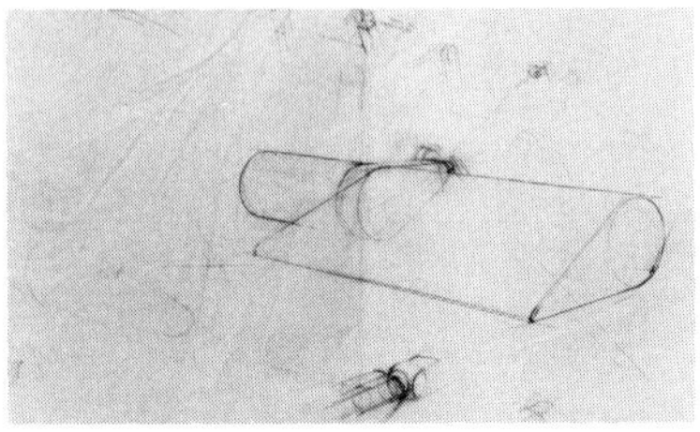

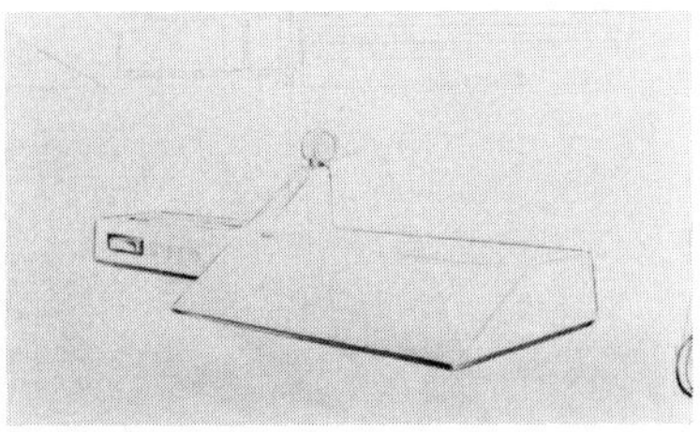

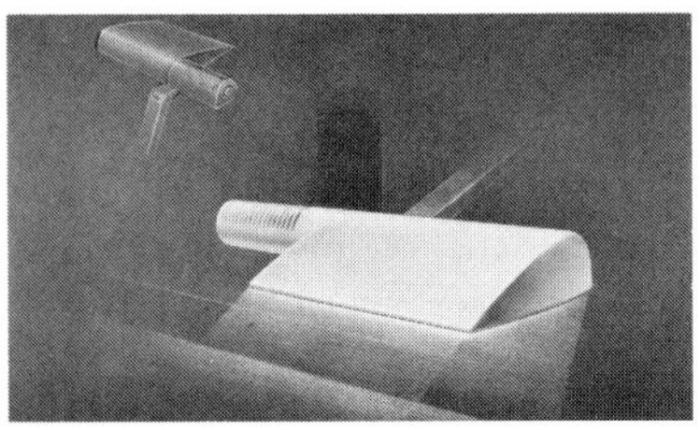

Sketches of early concepts.

The Design Problem: To design a light to meet the unique human factors and visual requirements of a new work environment.

This low energy, focused task light was designed to provide efficient, glare-free illumination of a small work area. The lamp's function had no traditional precedents.

Studies made of computer work stations showed that the use of display terminals and word processors presents unusual lighting requirements. Ambient lighting and light spilling from adjacent CRT work stations cause greater eye fatigue.

The designer states, "I wanted an absolute minimum of mass, a contemporary, light, horizontal look; the perception of an instrument rather than a light fixture, compatible with the appearance of adjacent electronics (1). The reflector shape evolved from the parabolic curve that was computer generated." The design uses a low voltage PL bulb, which when placed in the reflector, (2) distributes light evenly and eliminates any hot spots.

Detail of shield and double tube design.

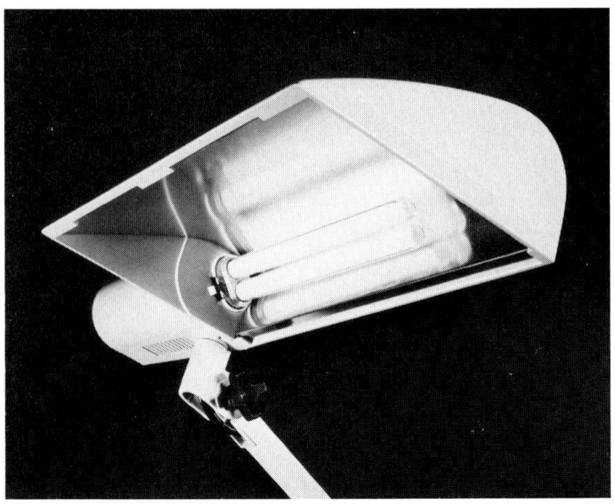

Task light designed especially for computer workstations.

Design consultants Designspring, Inc.
Creative director Will Goldschmidt
Director of engineering Irving Schaffer
Designer David Black
Manufacturer Ledu Corp.
Chief engineer (Scandinavia) Ralf Andersson
President Ledu USA John Stromberg

Evac + Chair® Emergency Descent 1981

How the unit handles a patient.

The braking-traction system of friction belts slow descent proportionately to the passenger's weight.

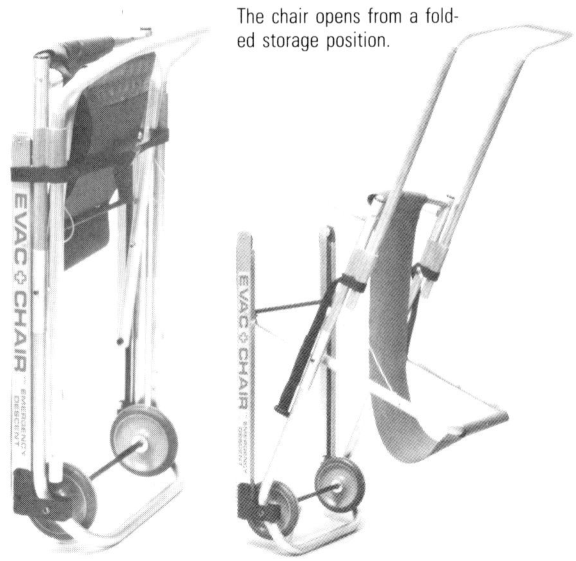

The chair opens from a folded storage position.

The Design Problem: To design a method of escape by stairway for the handicapped or disabled in an emergency situation.

Government statistics show that more than 9.4% of the white collar labor force has a chronic physical disability. The percentage is a great deal higher for the population as a whole. Neither count considers working women who are pregnant, the older worker and the temporarily disabled, or 12.5 million Americans at any one moment.

Until the EVAC + CHAIR was designed, there was no way for those who had problems with stairs to use them safely in an emergency without endangering themselves and others.

The designer began with a chair configuration and ended with a hybrid stretcher, cart and chair. Weighing less than 15 pounds, the unit needs no maintenance. It is designed for stand-by service in buildings, and for emergency use by ambulance, fire service and emergency medical teams.

The unit opens from its folded storage position in an instant. At the fire exit landing, the passenger evacuee sits in the chair, held in balance by the attendant who then guides the unique belted-traction carriages over the stairs at walking speed. The braking-traction system of friction belts slow descent proportionately to the weight of the passenger, and allows gravity and friction to do 90% of the carrying. David Egen believes that good design "can only flow from clearly defined and *true* necessities. Distinguishing the true from the less-than-true needs of the product is the crux."

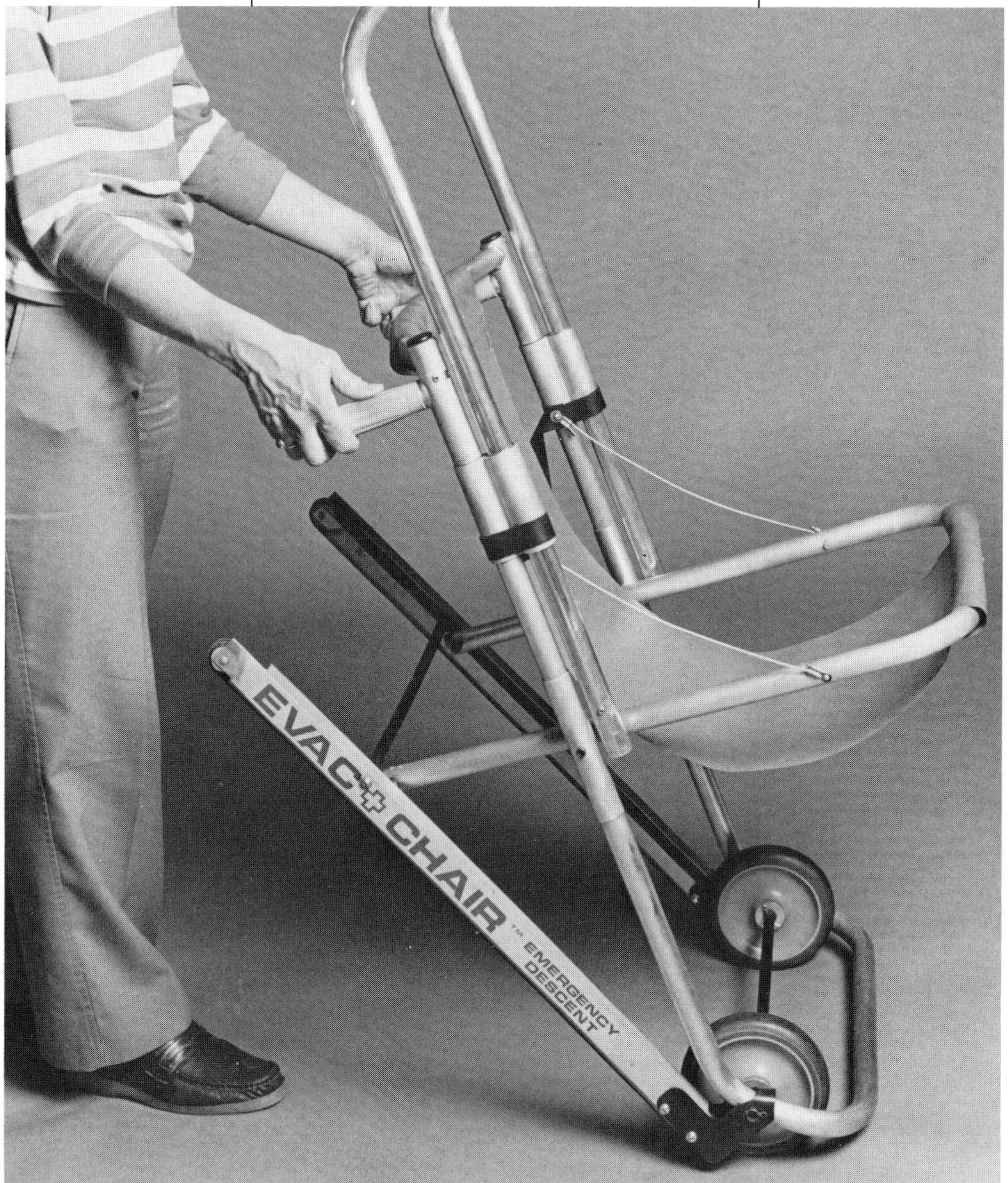

Manufacturer Evac + Chair Corp.
Designer David Egan

Knoll Advanced Operational Task Chair 1983

Niels Diffrient's concept drawings.

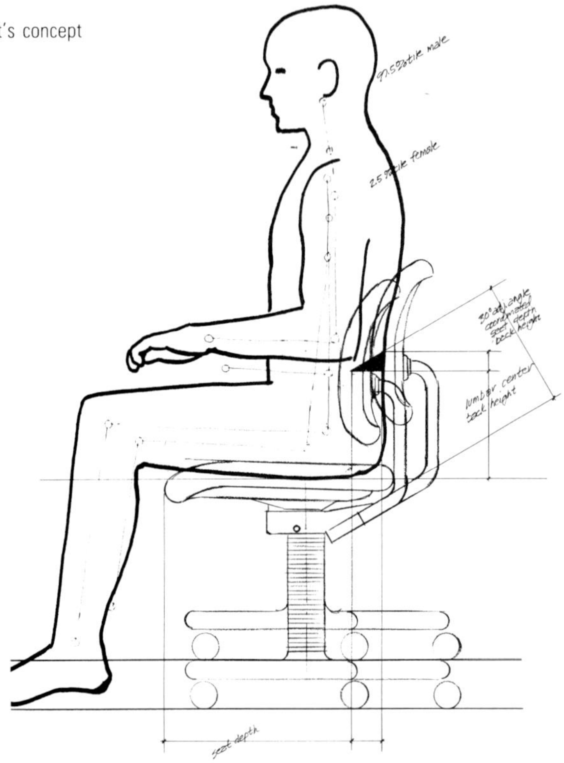

Adjustment geometry (coordinated seat and back adjustment to accomodate 95% of population).

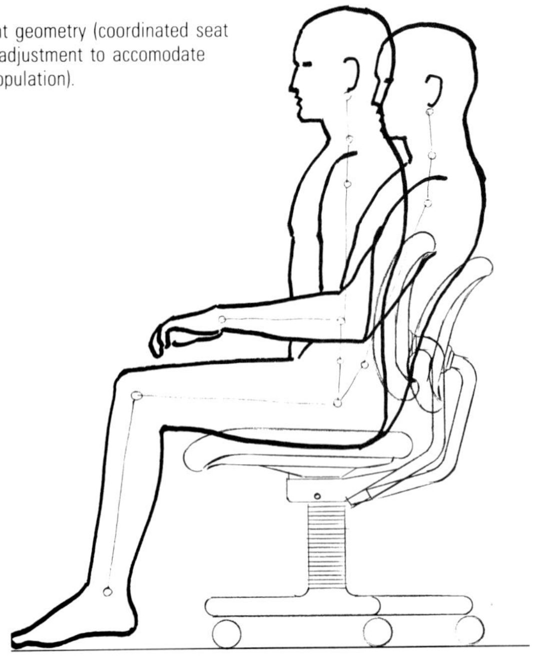

The Design Problem: To design a chair that allows the user to perform seated desk work with less strain.

The design is largely the by-product of research within the field of human factors engineering, in particular task seating, over a 28-year period. *Human-scale 1,2,3* (MIT Press, 1974), a study of human dimensions, was completed while Diffrient was associated with the Henry Dreyfuss design office.

According to Diffrient, "The turning point in any good functional design comes when form is not a product of preconceived style but is a direct out-growth of a more complete knowledge of product performance. Performance in this sense is determined by the creative analysis of the interrelation of product and user."

This chair has been designed to meet the needs of the automated office. The back height and seat depth are controlled by one adjustment lever, which is easily reached when seated; the back tilt adjustment is controlled by a fingertip lever. The chair tilts without the sitter's feet leaving the floor. Even when the chair is tilted backward, critical support for the neck and shoulders is provided.

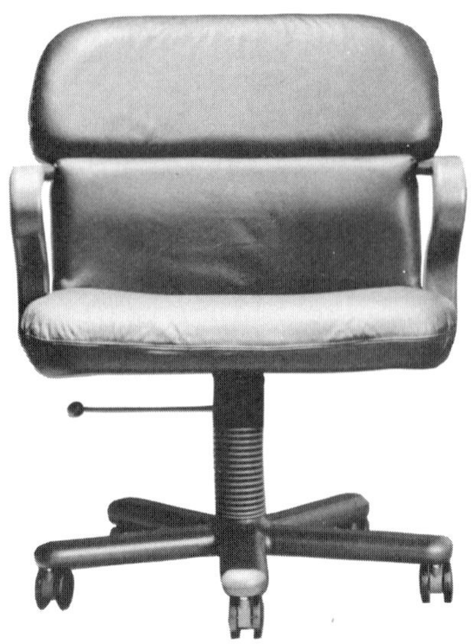

The comfort test device is used to take measurements and record comfort in seat sizes, back sizes, contours, cushion densities, spring forces, adjustment ranges, recline geometry and seat heights. The designer is seen here relaxing in the chair. After these studies are completed, functional appearance models will be made and a presentation given to the client.

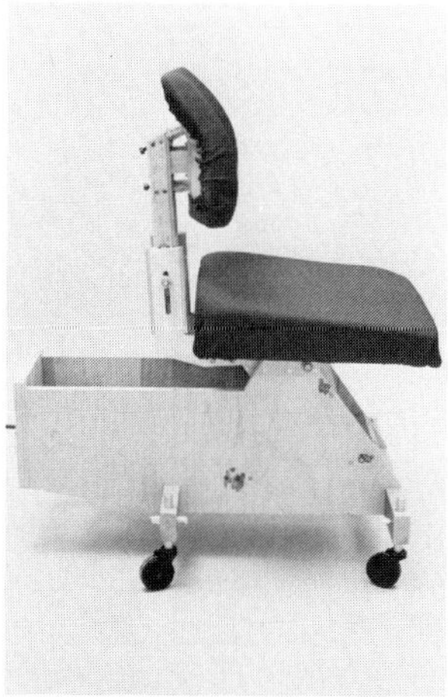

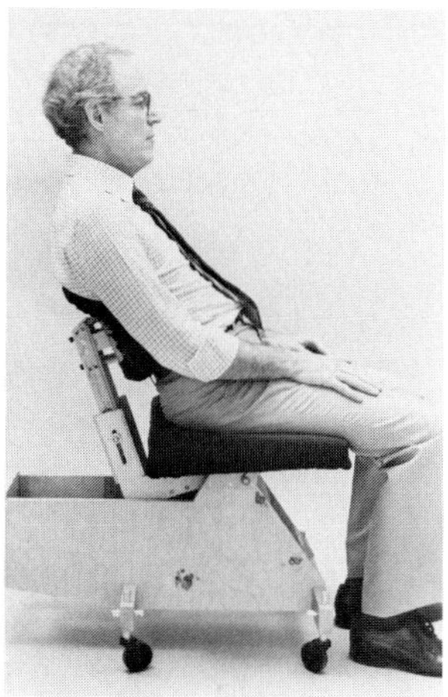

The designer testing the comfort and support factors in a rough mechanical version of the chair.

Designer Neils Diffrient
Manufacturer Knoll International, Inc.
Design development Tom Latone, Hy Zelkowitz

Designed to meet the needs of the automated office, the back height and seat depth are controlled by one adjustment lever, and the chair tilts without the sitter's feet leaving the floor.

Bibliography

Contemporary issues in industrial design

Busch, Akiko; *ID* editors. *Product Design,* New York, PBC International, 1984. More than 400 color photos of the best designs of the early 1980s.

Caplan, Ralph. *By Design,* New York, St. Martin's Press (hardcover), 1982, McGraw-Hill (paper), 1984.

Caplan, Ralph. *Design in America* (introduction), New York, McGraw-Hill, 1969.

Doblin, Jay. *One Hundred Great Product Designs,* New York, Van Nostrand Reinhold, 1970.

Evans, Helen Marie. *Man the Designer,* New York, Macmillan, 1973.

Heskett, John. *Industrial Design,* New York, Oxford University Press, 1980.

Hiesinger, Kathryn B. *Design Since 1945,* catalogue for the 1983–1984 exhibition at the Philadelphia Museum of Art, includes pieces by Max Brill, George Nelson, Niels Diffrient, Dieter Rams and Jack Lenor Larsen, Philadelphia, Philadelphia Museum of Art, 1983.

Hubel, Vello; Lussow, Diedra B. *Focus on Designing,* Toronto, Canada, Mc-Graw-Hill Ryerson, 1984. Excellent college-level textbook on design process.

Larabee, Eric; Vignelli, Massimo. *Knoll Design,* New York, Harry N. Abrams, 1981.

Nelson, George. *On Design,* New York, Whitney Library of Design, 1979.

Oakley, Mark. *Managing Product Design,* New York, John Wiley & Sons, 1984. How product design can be integrated into the modern corporation.

Papanek, Victor. *Design for the Real World,* 2nd ed., New York, Van Nostrand Reinhold (paperback), 1984.

Papanek, Victor. *Design for Human Scale,* New York, Van Nostrand Reinhold, 1982.

Pile, John. *Design, Purpose, Form and Meaning,* New York, W. W. Norton, 1979.

Pulos, Arthur. *Opportunities in Industrial Design,* Champagne, Illinois, V.G.M. Career Horisons, 1978.

Schutte, Thomas, ed. *The Uneasy Coalition: Design in Corporate America,* Philadelphia, lectures to Wharton School of Business.

Wallance, Don. *Shaping America's Products,* New York, Reinhold, 1956. Case histories of product designs.

Periodicals

ID (Industrial Design) The only U.S. publication in the field. First volume 1954; most recent volume, 32, for 1985. Includes 31 *ID Annual Design Reviews,* juried selections of outstanding product, environmental, and graphic designs. Design Publications, Inc., New York.

Innovation, The Journal of the Industrial Designers' Society of America. Theme issues published three times a year by IDSA, McLean, Va.

Exhibitions

Art and Industry, A Century of Design in Products We Use, Boilerhouse Project, Victoria & Albert Museum, London, U.K., January-March 1982.

Designs of the Year, Gallery 91, New York, NY, August-October 1985, the selections from the *1985 ID Annual Design Review.*

Design: The Problem Comes First, Danish Design Council, 1982.

Design Process Olivetti 1908–1978, Frederick S. Wight Gallery, University of California at Los Angeles, 1979.

Eight Decades of Vanguard Design, The Whitney Museum of American Art, New York, NY, fall 1985.

Italian ReEvolution: Design in Italian Society in the Eighties, LaJolla Museum of Contemporary Art, 1982.

Italy: The New Domestic Landscape, Achievements and Problems, Museum of Modern Art, New York, NY.

The Product of Design, The Design Process in America from Late 1950s to the Early 1980s, Katonah Gallery, Katonah, NY, Spring 1984; Nancy Hanks Center, Washington, D.C., August 1985, and other locations. (Basis for *American Design Classics)*

Twentieth Century Design USA, Albright-Knox Museum, Buffalo, NY, 1959. Excellent catalogue by William Friedman.

Design Since 1945, Philadelphia Museum of Art, October 1983-January 1984, Catalogue by Kathryn B. Hiesinger published by the Philadelphia Museum of Art, Philadelphia, 1983.

Design history

Banham, Reyner. *Theory of Design in the First Machine Age,* Cambridge, Mass. (hardcover), 1960 (paperback), 1980.

Bayley, Stephen. *In Good Shape: Style in Industrial Products 1900–1960,* New York, Van Nostrand Reinhold, 1979.

Bel Geddes, Norman. *Horisons,* New York, Little Brown, 1932. Personal reflections on design and its future.

Bush, Donald J. *The Streamlined Decade,* New York, George Brazilles, 1975.

Dreyfuss, Henry. *Designing for People,* New York, Simon & Schuster, 1955. Among the best descriptions of the design process by an eminent designer.

Drexler, Arthur; Daniel, Greta. *Introduction to Twentieth Century Design,* New York, Museum of Modern Art, 1959.

Ferebee, Ann. *A History of Design from the Victorian Era to the Present,* New York, Van Nostrand Reinhold, 1970.

Greif, Martin. *Depression Modern: the Thirties Style in America,* New York, Universe Books, 1975.

Katz, Sylvia. *Plastics: Common Objects, Classic Designs,* New York, Harry N. Abrams, 1984.

Kaufman, Edgar, Jr. *Introduction to Modern Design,* New York, Museum of Modern Art, 1950.

Loewy, Raymond. *The Design of Raymond Loewy,* Washington, D.C., exhibition catalogue, Renwick Gallery, 1976.

Loewy, Raymond. *Never Leave Well Enough Alone,* New York, Simon & Schuster, 1951.

Meikle, Jeffrey. *Twentieth Century Limited,* Philadelphia, Temple University Press, 1979.

Pevsner, Nikolaus. *The Sources of Modern Architecture and Design,* New York, Oxford University Press, 1968.

Pulos, Arthur. *American Design Ethic, A History of Industrial Design to 1940,* Cambridge, Mass., MIT Press, 1983.

Read, Herbert. *Art and Industry: the Principles of Industrial Design,* New York, Horizon Press (paperback), 1961.

Teague, Walter Dorwin. *Design This Day, The Technique of Order in the Machine Age,* New York, Harcourt Brace, 1940.

Nada Westerman and Joan Wessel, the authors, investigated recent design history for more than a year in preparation for curating the Product of Design exhibition, upon which *American Design Classics* is based. Both have long been students of contemporary art and design. Arthur J. Pulos, FIDSA, who wrote the introduction, has educated several generations of American designers while head of the department at Syracuse University. He is also the author of *American Design Ethic,* a benchmark history of design in America from the 18th Century to the 1940s, and is now preparing a second volume that will bring his design history up to the current period. Randolph McAusland, the editor, is publisher of *ID* magazine and the general manager of Design Publications, Inc. Joseph Kugielsky, who photographed most of the objects in the book, has had several shows and carried out assignments for *ID, The New York Times,* Stewart Tabori and Chang, and other publishers. Clive Jacobson, the designer, had his own design studio in South Africa and is a designer for Fulton + Partners.